PROGRESS IN PHARMACEUTICAL AND BIOMEDICAL ANALYSIS
*Series Editors*: C. M. Riley, A. F. Fell

VOLUME 3

# Development and Validation of Analytical Methods

**Related Elsevier Titles of Interest**

**BOOKS**

RILEY, LOUGH & WAINER: Pharmaceutical & Biomedical Applications of Liquid Chromatography
LUNTE & RADZIK: Pharmaceutical & Biomedical Applications of Capillary Electrophoresis
WONG & WHITESIDES: Enzymes in Synthetic Organic Chemistry
PELLITIER: Alkaloids, Chemical & Biological Perspectives Volume 9
Alkaloids, Chemical & Biological Perspectives Volume 10*

* In preparation

**JOURNALS**

BIOCHEMICAL PHARMACOLOGY
BIOORGANIC & MEDICINAL CHEMISTRY
BIOORGANIC & MEDICINAL CHEMISTRY LETTERS
JOURNAL OF PHARMACEUTICAL AND BIOMEDICAL ANALYSIS
CARBOHYDRATE RESEARCH
PHYTOCHEMISTRY

*Full details of all Elsevier Science publications/free specimen copy of any Elsevier Science journal are available on request from your nearest Elsevier Science office*

Front cover: Part of a figure illustrating the process nature of laboratory work and t relationships that must be included as a part of a validation program. (©Laboratory Automati Standards Foundation. Reproduced by permission of the copyright holder)

# Development and Validation of Analytical Methods

*Edited by*

**Christopher M. Riley**
*Analytical Research and Development*
*DuPont Merck Pharmaceutical Company*
*Wilmington, DE 19880, USA*
*and*
*Department of Pharmaceutical Chemistry*
*University of Kansas,*
*Lawrence, KS 66045, USA*

*and*

**Thomas W. Rosanske**
*Analytical Chemistry Department*
*Hoechst Marion Roussel, Inc.,*
*Kansas City, MO 64134, USA*

PERGAMON

U.K. Elsevier Science Ltd, The Boulevard, Langford Lane,
Kidlington, Oxford OX5 1GB, U.K.
U.S.A. Elsevier Science Inc., 660 White Plains Road,
Tarrytown, New York 10591-5153, U.S.A.
JAPAN Elsevier Science Japan, Tsunashima Building Annex,
3-20-12 Yushima, Bunkyo-ku, Tokyo 113, Japan







First Edition 1996

**Library of Congress Catalogue in Publication Data**

Pharmaceutical and biomedial analysis: development and validation of analytic methods / edited by Christopher M. Riley and Thomas W. Rosanske.
-- 1st ed.
p. cm. -- (Progress in pharmaceutical and biomedial analysis; v. 3)
Includes index.
1. Drugs--Analysis--Methodology. I. Riley, Christopher M. II. Rosanske, Thom W. III. Series
[DNLM: 1. Chemistry, Pharmaceutical--methods. 2. Drugs--analysis.
W1 PR677LM v. 3 1996 / QV 25 P5353 1996]
RS189.P442 1996
615' . 1901--dc20
DNLM/DLC
for Library of Congress 96-33800
CIP

**British Library Cataloguing in Publication Data**

A catalogue record for this book is available from the British Library

ISBN 0 08 042792 8

*Printed and Bound in Great Britain by Biddles (Printers) Ltd, Guildford*

## Table of Contents

## List of Contributors

**Cynthia K. Brown**, Department of Analytical Chemistry, Hoechst Marion Roussel Inc., Marion Park Drive, Kansas City, MO 64134, USA

**Cathy L. Burgess, esquire**, Winston & Strawn, 1400 L Street NW, Washington, DC 20005, USA

**Ian E. Davidson**, Department of Analytical Chemistry, Hoechst Marion Roussel Inc., Marion Park Drive, Kanas City, MO 64134, USA

**Richard J. Davis**, Formerly Mid-Atlantic Office of the Food and Drug Administration, Philadelphia, PA, USA; present address Quality Assurance/Regulatory Compliance, DuPont Merck Pharmaceutical Company, DuPont Merck Plaza, Wilmington, DE 19880-0722, USA

**Paul K. Hovsepian**, Analytical Research and Development, DuPont Merck Pharmaceutical Company, Wilmington, DE 19880, USA

**Joseph G. Liscouski**, Laboratory Automation Standards Foundation, PO Box 38, Groton, MA 01450, USA

**Eugene McGonigle,** Physical and Analytical Chemistry, Research and Development, Schering-Plough Research Institute, Kenilworth, NJ 07033, USA

**Christopher M. Riley**, Analytical Research and Development, DuPont Merck Pharmaceutical Company, Wilmington, DE 19880, USA; and Pharmaceutical Chemistry Department, University of Kansas, Lawrence, KS 66045, USA

**Thomas W. Rosanske**, Department of Analytical Chemistry, Hoechst Marion Roussel Inc., Marion Park Drive, Kanas City, MO 64134, USA

**Thomas M. Rossi**, Analytical Research and Development Department, The R.W. Johnson Pharmaceutical Research Institute, Raritan, NJ 08869, USA

**Ralph R. Ryall**, Analytical Research and Development Department, The R.W. Johnson Pharmaceutical Research Institute, Raritan, NJ 08869, USA

**Krzysztof A. Selinger**, Clinical Pharmacology Department, Glaxo Wellcome Inc., Five Moore Drive, Research Triangle Park, NC 27709, USA

**G. Susan Srivatsa**, Development Analytical Chemistry, Isis Pharmaceuticals, 2292 Faraday Avenue, Carlsbad, CA 92008, USA

**Julie J. Tomlinson**, Bioanalytical Automation and Robotics, Clinical Pharmacology Department, Glaxo Wellcome Inc., Five Moore Drive, Research Triangle Park, NC 27709, USA

# Introduction

The need to validate an analytical or bioanalytical method is encountered by analysts in the pharmaceutical industry on an almost daily basis, because adequately validated methods are a necessity for approvable regulatory filings. What constitutes a validated method, however, is subject to analyst interpretation because there is no universally accepted industry practice for assay validation. This book is intended to serve as a guide to the analyst in terms of the issues and parameters that must be considered in the development and validation of analytical methods.

The scope of this book was expanded considerably from our original intent of considering only the validation of analytical methods. As we began the process of planning the layout of the book, it became clear that factors such as method development, data acquisition systems and regulatory considerations were all closely interrelated with validation and that a text on validation would not be complete without rather detailed discussions of all these components.

The book is divided into three parts. Part One, comprising two chapters, looks at some of the basic concepts of method validation. Chapter 1 discusses the general concept of validation and its role in the process of transferring methods from laboratory to laboratory. Chapter 2 looks at some of the critical parameters included in a validation program and the various statistical treatments given to these parameters.

Part Two (Chapters 3, 4 and 5) of the book focuses on the regulatory perspective of analytical validation. Chapter 3 discusses in some detail how validation is treated by various regulatory agencies around the world, including the United States, Canada, the European Community, Australia and Japan. This chapter also discusses the International Conference on Harmonization (ICH) treatment of assay validation. Chapters 4 and 5 cover the issues and various perspectives of the recent United States vs. Barr Laboratories Inc. case involving the retesting of samples.

Part Three (Chapters 6 - 12) covers the development and validation of various analytical components of the pharmaceutical product development process. This part of the book contains specific chapters dedicated to bulk drug substances and finished products, dissolution testing, robotics and automated systems, biotechnology related products, materials presented in biological matrices, cleaning procedures, and data acquisition systems. Each chapter goes into some detail describing the critical development and related validation considerations for each topic.

This book is not intended to be a practical description of the analytical validation process, but more of a guide to the critical parameters and considerations that must be attended to in an analytical development program. Despite the existence of numerous guidelines including the recent attempts by the ICH, the practical part of assay validation will always remain, to a certain extent, a matter of the personal preference of the analyst or company. Nevertheless, this book brings together the perspectives of several experts having extensive experience in different capacities in the pharmaceutical industry in an attempt to bring some consistency to analytical method development and validation.

The initial concepts for this book were developed while one of us (CMR) was on sabbatical leave from the University of Kansas at Hoechst Marion Roussel in Kansas City. We wish to express our gratitude to Dr. Michael Baltezor (Senior Director, Analytical Chemistry, Hoechst Marion Roussel) for providing us with access to resources in his department and for the opportunity to collaborate on this book.

Christopher M. Riley
Thomas W. Rosanske
January 2, 1996

# Part One: Basic Concepts

# Chapter 1

# Assay Validation and Inter-laboratory Transfer

*Eugene McGonigle*

## 1.1 Introduction

Although the process of transferring an analytical method from one laboratory to another uses some of the principles and components of validation, validation and method transfer should be regarded as two distinct processes. The purpose of this chapter is to summarize the relationship between method transfer and validation, as well as to introduce some of the terminology that will be discussed later in this book.

Validation may be viewed as the establishment of an experimental database that certifies an analytical method performs in the manner for which it was intended and is the responsibility of the method-development laboratory. Method transfer, on the other hand, is the introduction of a validated method into a designated laboratory so that it can be used in the same capacity for which it was originally developed. Ordinarily, the method-transfer process should be the responsibility of the designated laboratory that will use the previously validated method. However, successful method transfer relies upon close cooperation and communication between the two laboratories.

## 1.2 Functional definitions

At this point, the introduction of some functional definitions is necessary to distinguish the responsibilities of the laboratory that develops a method from those of the laboratory who will use the method. The "gap" between the two laboratories is bridged by the method-transfer process.

In a pharmaceutical-industrial setting, the Analytical Research and Development (ARD) group usually provides validated analytical methods. Consistent with this, when a method is submitted to regulatory agency, a Method Validation Report will be provided. The method and the supporting data in the report are reviewed by both manufacturing and control reviewing chemists and one or more validation laboratories at the agency. The assessment by the validation laboratory is referred to as the "Validation of the Analytical Procedure". While use of the term validation in all the above situations is correct, it is not necessarily correct to state that a method used by the Quality Control (QC) Department to release commercial goods has been validated by that department. It may be correct if the QC Department developed the procedure, or if, for some reason, they choose to repeat the original method validation experiments.

Although the ARD group normally develops the original method, the real answer to the questions "who validates analytical methods" and "who transfers them" depends upon the circumstances. For example, analytical methods can either be developed by or transferred to: the laboratories, QC laboratories or third parties such as contract laboratories, depending upon a company's organizational structure and their objectives at a particular point in time. At most pharmaceutical companies the ARD laboratories are responsible for the development of analytical methods and their charter is to provide appropriate test methods, specifications and stability data to support all domestic and international registrations of new pharmaceutical entities and related dosage forms. Clearly, analytical methods for newly registered products require validation reports. However, it is equally probable that the QC Department may modify existing methods or develop improved procedures as new technology emerges. For example, the QC department may undertake developing a more rapid and efficient chromatographic separation rendering a particular analysis more cost effective.

Analytical methods may also be transferred to either an ARD laboratory or a QC laboratory from outside sources. This is likely to occur more frequently in the future as companies increase their activity in the area of licensing and acquisition of new products or, due to limited internal resources, sub-contract an increasingly larger fraction of their method development activities to contract laboratories. These possibilities reinforce the necessary distinction between method validation and method transfer since both ARD and QC laboratories can be involved in both processes. The experimental rationale and Standard Operating Procedures (SOPs) for method validation and method transfer are different in both their objectives and their utilization[1].

## 1.2 Essential principles of method transfer

There are five essential principles, which if followed, will ensure successful method transfer: documentation, communication, acceptance criteria, implementation, and method modification and revalidation.

### 1.2.1 Documentation

The first essential principle of successful method transfer requires that the development laboratory provide the designated laboratory that will utilize the method with three essential elements, which demonstrate collectively that an analytical method is fully validated and ready for transfer. These include:

**a) a written procedure**, which contains a detailed, step-by-step description of the manipulations, specific reagents, equipment, instrument settings and other critical parameters. Each step in the procedure should be an explicit instruction allowing only one possible interpretation. This element of detail, from which there should be no essential deviation, most distinguishes method transfer from method validation or, the more general category under which method transfer is sometime placed, "technology transfer". The procedure must have a unique, cataloged identification so there can be no doubt that the correct procedure has been provided and received.

**b) a method-validation report** (see also sec. 1.3), which should include both the experimental design and the data that justify the

---

[1] Editors' note: Although this chapter is concerned mainly with the validation and transfer of methods developed for the characterization of the drug substance and related pharmaceutical products, much of the discussion is equally applicable to validated methods for the analysis of drugs and metabolites in biological samples.

conclusion that the analytical method, as written, performs as intended; and

c) **system suitability criteria**, which define the minimum acceptable performance criteria prior to each analysis.

### 1.2.2 Communication

As part of the FDA's Pre-approval Inspection (PAI) process, more and more emphasis is being placed on verifying that the official (*sic*) QC method is consistent with that included in the New Drug Application (NDA) dossier. Therefore, the second principle of successful method transfer is that ARD and QC staffs should meet **before** transfer to discuss all relevant, practical aspects of the method, particularly the manipulative steps. Ideally, these discussions should be initiated before validation is complete. In this fashion, it may be possible to incorporate any desirable modifications into the ARD validation report prior to registration. If such modifications constitute an equivalent alternative, early interdepartmental communications allow time to generate the data necessary to justify the alternative or to prepare a rationale why additional data are unnecessary. This also allows for preparing an Alternative Method Validation Report to be written and included in the registration package. Typical examples of equivalent alternatives include: changes in sample preparation automation requirements and the availability, suitability or cost-effectiveness of required reagents and equipment.

### 1.2.3 Acceptance criteria

The third principle requires that the designated laboratory be responsible for issuing and following SOPs that define their criteria for accepting an analytical method. Data generated in accordance with those SOPs form the basis for the Method Transfer Report, which should issue from the designated laboratory. This is important because the designated laboratory must ultimately assume responsibility for data and conclusions resulting from use of the method. As with most SOPs, it is both probable and acceptable for laboratories to have unique criteria for defining a method as acceptable. Some examples where different acceptance criteria may be used are: different statistical approaches for data evaluation, differences in the desired linear range, and different schemes for evaluating operator-to-operator or day-to-day variability.

### 1.2.4 Implementating the method as written and validated

This principle is the most important. The designated laboratory must follow the procedure, as written, to ensure that the method is supported by the database to be included in the Method Validation Report. Adherence to this principle will preclude the need to add additional data at the time of method transfer or, even worse, repeating the validation experiments. This again emphasizes the importance of effective communications between ARD and QC staffs.

### 1.2.5 Method modification and revalidation

Finally, if significant modifications to a method are incorporated at the time of transfer, revalidation may be necessary to ensure that the modifications have not invalidated previous, conclusive data in the Method Validation Report. Obviously, not all changes to the method require revalidation. Mention was made in sec. 1.2.2 of the potential advantages of discussing/incorporating modifications to an analytical method before method transfer. With the possible exception of automation, the examples of method modification given above represent method revisions and, therefore, may require revalidation of the method and, more importantly, a new and unique catalogue revision . The latter is important to ensure that there is no confusion over which version of the method has been used or is being employed for product analysis or which version is associated with a particular Method Validation or Method Transfer Report.

The following are some examples of modifications and subsequent aspects of method development that require revalidation:

**a) automation** The most common example of automation is the use of robotics (see also Chapter 7). If the robot repeats, identically, all manipulations of the previous manual method, only the precision experiments need be repeated. Since this might occur in a laboratory after method transfer, a statistical comparison of the precision of the manual and automated methods may be conducted on the same population of a representative product. Obviously, this assumes that the robotic system has been appropriately calibrated.

**b) sample preparation** One example of a change in sample preparation would be the desire to use whole tablets instead of ground-tablet composites. Another example might be a modification in which the solvent:solvent or solvent:solid ratios are changed in the extraction step. In these examples, accuracy, precision, linearity and range may be affected so the validation of these parameters should be repeated using both synthetic preparations of the drug product components and representative product samples.

**c) dilutions** If the solvent ratios remain identical to those in the original method and the analytical concentrations are also the same, only the precision is likely to be affected and those experiments should be repeated. Analysis of a representative product would be adequate for this purpose. In contrast, if the solvent aliquots are substantially different or the analytical solvent ratios differ in the dilution step(s), linearity and precision should be reassessed. Synthetic preparations of the drug product components should be utilized to re-establish comparable analytical parameters. Testing a representative product is probably not necessary but most laboratories tend to include this. If the proposed dilution change effects a different analytical concentration the analyst should also consider re-evaluating the specificity and the resolution (for chromatographic methods).

**d) alternative chromatographic columns** Substituting an alternative chromatographic column always raises the possibility of a change in specificity and resolution as well as the quantitative aspects of the method. Therefore, modifications of this type require that all method validation parameters to be reassessed, that is specificity(resolution), linearity, range, accuracy, precision and limit of quantitation (LOQ).

## 1.3 Method Validation Report

The preceding sections emphasize the importance of distinguishing between method validation and method transfer. Method validation establishes the scientific qualification of a specific analytical method and details of this are included in the Method

Validation Report. Transfer of a validated method is governed by the SOP established by the designated laboratory, which defines their acceptable performance criteria. Thus, the Method Validation Report is a pivotal document for any regulatory submission because it forms the basis for scientific qualification of the method. Consequently, it is appropriate to review certain aspects of the Method Validation Report, which relate directly to the method-transfer process and thereby qualify acceptable performance in the designated laboratory. Readers are referred to the other chapters in this book for more complete definitions and descriptions of the essential elements of assay validation.

### 1.3.1 Specificity (selectivity)

For a stability indicating assay of an intact drug, specificity (or selectivity)[2] defines the ability of the method to measure the analyte to the exclusion of relevant components, which might interfere. Experiments to establish method specificity include evaluating formulation matrix components (e.g. a placebo) and any known related compounds such as synthesis-related impurities and degradation products. Other less relevant compounds such as metabolites or isomers, which might help to define the limits of a method's resolution may also be evaluated. A similar assessment is repeated after stressing the drug to accelerate degradation under the influence of heat, light, oxidation, and acid and base hydrolysis.

### 1.3.2 Chromatographic parameters/system suitability

For chromatographic methods a number of quantitative features should be defined. These parameters are ultimately used as the minimum standards of performance in system suitability tests. The resolution of a crucial pair (or pairs) of peaks in the chromatogram defines minimum separation requirement(s). Thus, the minimum resolution factor in the system suitability tests is generally used in conjunction with the column efficiency (number of theoretical plates) and the tailing factor.

---

[2] See Chapter 2 for a detailed discussion of the differences between selectivity and specificity

### 1.3.3 Linearity

Linearity defines the analytical response as a function of analyte concentration and range prescribes a region over which acceptable linearity, precision and accuracy are achieved. While many analysts prefer to use a broad range, quantitative measurements for pharmaceutical assays (c.f. bioanalytical methods) are made over a range that minimally encompasses 80%, 100% and 120% of the analytical concentration prescribed in the method, recognizing this can be different than product "label claim". Either a sample of the drug substance or the reference standard may be used for these experiments.

### 1.3.4 Accuracy (recovery)

Recovery of the analyte of interest from a given matrix can be used as a measure of the accuracy or the bias of the method. The same range of concentrations as employed in the linearity studies is used. That is, the linearity experiment is repeated in the presence of matrix constituents; however, incorporation of impurities and degradation products may also be appropriate. It is important to note that, for the purposes of this discussion, recovery experiments of this type (while commonly employed) are limited to the detection of positive matrix interferences or adsorption effects and are not true "recovery" experiments. This is especially true of analytical methods for solid dosage forms. Furthermore, the method of additions can mask positive or negative bias and should only be used as a last resort.

### 1.3.5 Precision

Precision quantifies the variability of an analytical result as a function of operator, method manipulations and day-to-day environment. Statistical analysis of data generated to demonstrate assay precision is essential. For efficiency, analysts can use both the linearity and the recovery data for the statistical assessment. In addition, it is prudent to include an additional tier of comparative analyses of the sample of a representative product, usually a minimum of ten replicates. All the data to this point will have been generated by the development laboratory chemists. From the perspective of the designated laboratory, these validation data should be viewed as intra-laboratory data, even though more than one development chemist or more than one development laboratory may have been involved in generating the precision data.

## 1.4 The Inter-laboratory Qualification (ILQ) Process

Clearly the method-transfer process should involve more than simply the designated laboratory obtaining the "expected" result after analyzing a sample of representative product. This alone will not assure consistent performance of the method over time and may actually mask erroneous results arising from compensating errors. Furthermore, analysts in the pharmaceutical industry must be prepared answer the question: *"Why do you always accept good results without challenge and question only what you conclude to be incorrect assays?"* A good QC laboratory must be able to recognize rapidly, and with confidence, a method that is out of control, independent of the assay result, before a product is released. There are a number of ways to achieve this level of performance, begining with a sound method-transfer protocol.

Each laboratory involved in the method transfer process should define, independently, an experimental protocol to be followed for every method transferred. However, it is obvious that the most efficient way is to take advantage of the scientific database already established and included in the Method Validation Report. First, the designated laboratory should confirm the linearity and recovery for the analyte alone and in the presence of the known product components. Designing the experimental protocol for the ILQ so that it resembles, as much as possible, that which was carried out by the development laboratory provides two advantages:

a) it allows comparison of resultant raw data and calculated results with those already in the Method Validation Report; and

b) assuming the analytical method remains essentially unchanged during transfer, the continuity of experimental design and resultant data allow the validation and the transfer reports to be reviewed by a regulatory agency as a complementary package.

Both the method development laboratory and the designated laboratory should test a common sample population that should be representative of the intended product. Comparison of these data provide an additional level of intra-laboratory information as well as forming the basis of the inter-laboratory qualification (ILQ) process. This simple experimental design generally allows bias or imprecision to be traced to an instrument in one of the laboratories, the method itself or a specific operator, the reason being that both laboratories will have analyzed both "analytically prepared" and "representative product" samples. Bias or imprecision associated with the assay of the former are clearly method related. Anomalous results associated only with latter usually indicate a problem with the conduct of the method in the designated laboratory, which under normal circumstances can be corrected by incorporating additional instructions into the method or by further training.

The results and conclusions of these experiments are summarized in the Method Transfer Report. Since the overall objective of this report should be documentation that the method is acceptable, it is the responsibility of the designated laboratory. The Method Transfer Report should remain in the files of the designated laboratory along with the Method Validation Report to support subsequent audits.

## 1.5 Conclusions

Method validation and method transfer are distinct processes. Method validation certifies that the method performs in the manner for which it was developed and is the responsibility of the method development laboratory. Experimental rationale, the supporting database and conclusions are summarized in the Method Validation Report. Method transfer certifies that an analytical method is acceptable to the designated laboratory once it has generated data demonstrating the method performs in the same capacity for which it was validated. Accordingly, method transfer criteria should be based on the SOPs, which are unique to the designated laboratory. While selected in accordance with minimum scientific criteria, method transfer SOPs should be consistent with the unique requirements of the individual laboratories, which may not (and need not) be identical in all laboratories. Finally, supporting data and conclusions for method transfer should be documented in a Method Transfer Report prepared by and retained in the designated laboratory.

## Chapter 2

# Statistical Parameters and Analytical Figures of Merit

*Christopher M. Riley*

## 2.1 Introduction

Validation of an analytical method is primarily concerned with the identification of the sources and the subsequent quantification of the potential errors in the method. Although basic assay validation uses relatively simple statistical approaches, Miller and Miller [1] claim that "......*many highly competent scientists are woefully ignorant of even the most elementary statistical methods. It is even more astonishing that analytical chemists, who practise one of the most quantitative of all sciences are no more immune than others to this dangerous, but entirely curable, affliction.*" Thus, no text on the subject of validation would be complete without a discussion of the sources and types of error that may occur in the course of an analytical determination. Readers wishing to learn more about the application of statistics to analytical chemistry are referred to several excellent texts on the subject [2–5]. The recently revised edition of "Statistics for Analytical Chemistry" by Miller and Miller [2] is highly recommended. Three types of error may be encountered in an analytical measurement: gross, systematic and random; and the analyst should be able to distinguish between each type.

### 2.1.1 Gross errors

Gross errors are simply defined as errors that are so serious that they require termination of the experiment. Such errors might include loss or contamination of a sample, omission of critical reagents, or instrument failure. Occasional gross errors are inevitable and ordinarily they are easily recognized. However, the recent legal action brought by the Food and Drug Administration (FDA) against Barr Laboratories, which focused on Barr Laboratories' procedures for retesting samples that were out of specification, highlights the importance of defining what constitutes a failure, as well as the procedures for failure investigations and for retesting of samples. The case of the United States *versus* Barr Laboratories [6] will be discussed in more detail in Chapter 4.

### 2.1.2 Systematic errors

Systematic errors (or determinate errors) affect the accuracy of an analytical measurement causing all the results to be in error in the same sense [7]. The origin of systematic errors can usually be traced to errors of the analyst, instrumental errors (i.e. errors in calibration of the instrument), errors introduced by the reagent (e.g. side reactions or incomplete reactions) or some combination of the above. Systematic errors can bias the results so that they will be all too high or all too low. The systematic error associated with an analytical measurement can be defined by the bias, which may be positive or negative, or the accuracy, which may range between 0 (completely inaccurate) and 100 (completely accurate).

$$\text{Bias} = \left(\frac{\mu - \bar{x}}{\mu}\right) \cdot 100 \qquad (2.1)$$

$$\text{Accuracy} = \left(\frac{\bar{x}}{\mu}\right) \cdot 100 \qquad (2.2)$$

where:

$$\bar{x} = \frac{\sum_{i=1}^{i=n} x_i}{n} \qquad (2.3)$$

and $\mu$ is the true mean (or the population mean), $\bar{x}$ is the mean or the average of n measurements made on a sample taken to be representative of the total population and $x_i$ is the value of the ith measurement.

### 2.1.3 Random errors

Random errors (or indeterminate errors) result from uncontrolled variables in the measurement conditions, which cause individual results or measurements to fall on either side of the true mean, $\mu$. Random errors can be reduced but can never be eliminated. However, they are amenable to statistical analysis. If the individual results are normally distributed about the population mean, and there are no systematic errors (c.f. Sec. 2.1.2), the probability ($y_i$) of obtaining a value $x_i$ is given by:

$$y_i = \frac{e^{\left\{\frac{-(x_i-\mu)^2}{2\sigma^2}\right\}}}{\sigma\sqrt{2\pi}} \tag{2.4}$$

where $\sigma$ is the standard deviation of the population (eq. 2.5). The larger the value of the standard deviation, the greater the spread of the data about the mean (Fig. 2.1).

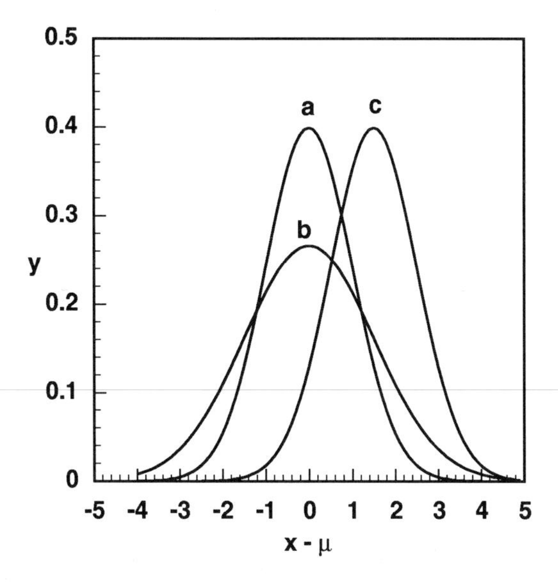


**Figure 2.1**
Normal distribution of experimental data with different values for the means ($\mu$) and the standard deviations ($\sigma$) plotted according to eq. 2.4. Key: $\mu_a = \mu_b < \mu_c$; $\sigma_a = \sigma_c < \sigma_b$

$$\sigma = \sqrt{\frac{\sum_{i=1}^{i=n} (x_i - \bar{x})^2}{n}} \tag{2.5}$$

It is possible for the distribution of data to be asymmetrically distributed about the mean as shown in Fig. 2.2. Nevertheless the methods for the analysis of data are essentially the same, irrespective of whether they are normally or asymmetrically distributed. Equation 2.4 predicts and Fig. 2.3 shows that approximately 68% of the values lie within $\pm 1\sigma$ of the mean, approximately 95% of the values lie within $\pm 2\sigma$ of the mean and approximately 99.7% of the values lie within $\pm 3\sigma$ of the mean. Such information allows the description of confidence intervals about the sample mean.

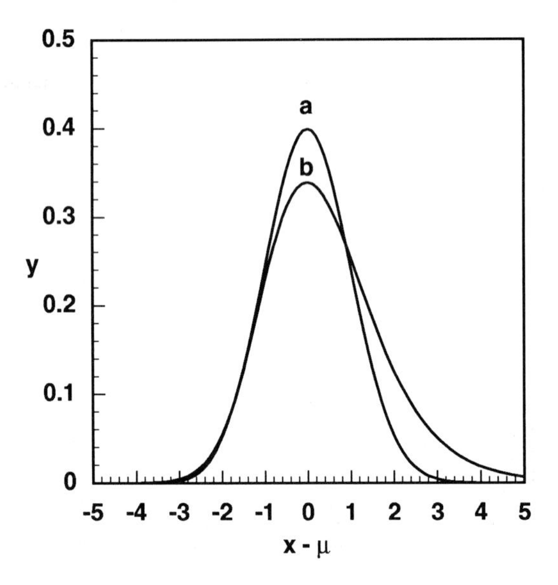


**Figure 2.2**
Comparison of (a) normal and (b) asymmetric (skewed) distribution of data

Because the true standard deviation of the population (σ) can be determined only if a very large number of measurements are made, an estimate of the standard deviation of an analytical method is usually made with a much smaller set of measurements that constitutes a sample of the all the measurements that could be made. In this case the standard deviation of the sample is given by:

$$s = \sqrt{\frac{\sum_{i=1}^{i=n} (x_i - \bar{x})^2}{n - 1}} \tag{2.6}$$

If the number of measurements is small then the standard deviation will be underestimated if eq 2.5 is used instead of eq. 2.6.

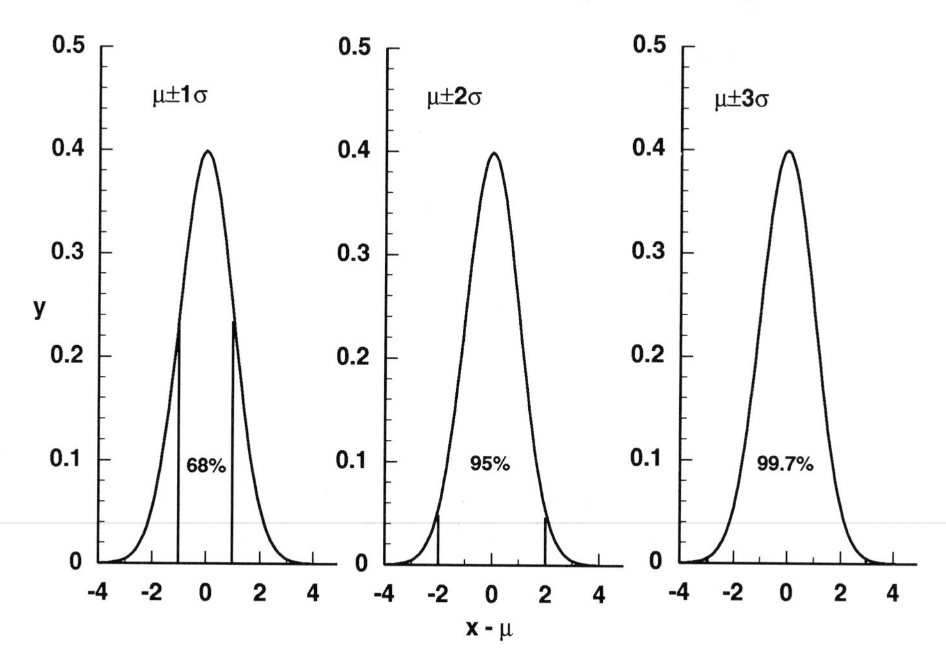


**Figure 2.3**
Relationship between normal distribution of data and standard deviation

## 2.2 Validation parameters

The primary statistical parameters that validate an analytical method are the accuracy and precision. Although the validity of experimental data may be defined primarily by the accuracy and precision of the analytical method used to generate those data, the use of these parameters alone is generally considered inadequate and supplementary experiments are necessary for the validation of a new method to be complete.

The following sections summarize the various parameters that have been described for the validation of quantitative analytical methods. The procedures used for the validation of qualitative methods are generally less involved and are usually concerned mainly with the establishment of selectivity or specificity and ruggedness.

### 2.2.1 Accuracy and precision

The precision and accuracy of analytical methods are described in a quantitative fashion by the use of relative errors. One example of relative error is the accuracy (eq. 2.2), which describes the deviation from the expected result. The relative error term usually used to describe precision is the relative standard deviation (RSD): [1]

$$\mathrm{RSD} = \frac{s}{\overline{x}} \cdot 100 \qquad (2.7)$$

The precision of the analytical system and the precision of the method are generally defined separately. This is because the former provides information on the errors associated with the instrumentation and the latter provides information on the complete method. The difference between the two ordinarily arises from errors associated with sample preparation. For example in liquid chromatography, the system precision may be determined from the RSD of repetitive injections (n=5 or 6) of the same solution. In contrast to the system precision, the method precision is determined by repetitive analysis (n=5 or 6) of a single homogeneous sample. For

---

[1] The relative standard deviation is also known as the coefficient of variation (CV).

example the system RSD for an LC method may be assumed to be a function of the random errors arising from the column, the injector, the detector and the integration device. Reasonable estimates of the RSDs attributable to these components might be:

$$RSD_{column} = 0.1\%$$

$$RSD_{injector} = 0.5\%$$

$$RSD_{detector} = 0.3\%$$

$$RSD_{integrator} = 0.1\%$$

in which case the system RSD is given by:

$$\begin{aligned} RSD_{system} &= \sqrt{0.1^2 + 0.5^2 + 0.3^2 + 0.1^2} \\ &= 0.60\% \end{aligned} \tag{2.8}$$

The error attributable to sample preparation will vary considerably depending on the number of steps, the complexity of each step and the concentration of the analyte of interest. For a simple LC method for the determination of the purity of a drug substance, the sample preparation might be relatively straightforward involving weighing the drug, dissolving in a suitable solvent and adjusting to volume. In this case, the RSD for the sample preparation step ($RSD_{prep}$) might be approximately 1%. The method RSD is then given by:

$$\begin{aligned} RSD_{method} &= \sqrt{0.1^2 + 0.5^2 + 0.3^2 + 0.1^2 + 1.0^2} \\ &= 1.17\% \end{aligned} \tag{2.9}$$

Equations 2.8 and 2.9 illustrate a very important point that the overall random error associated with a particular determination or method is dominated by the least precise step or component, so measures designed to improve the precision should always be directed towards improving the step or component having the highest degree error. Thus, for the hypothetical example shown here, the system precision in LC is governed by the precision of the injection device and the method precision is governed by the complexity of the sample preparation procedure.

The distinction may be made between the within-run precision (also refered to as the within-day or intra-laboratory precision) and the between-day (between-run or inter-laboratory) precision of an analytical method. These two terms are sometimes referred to as the repeatability and the reproducibility of the method, respectively.

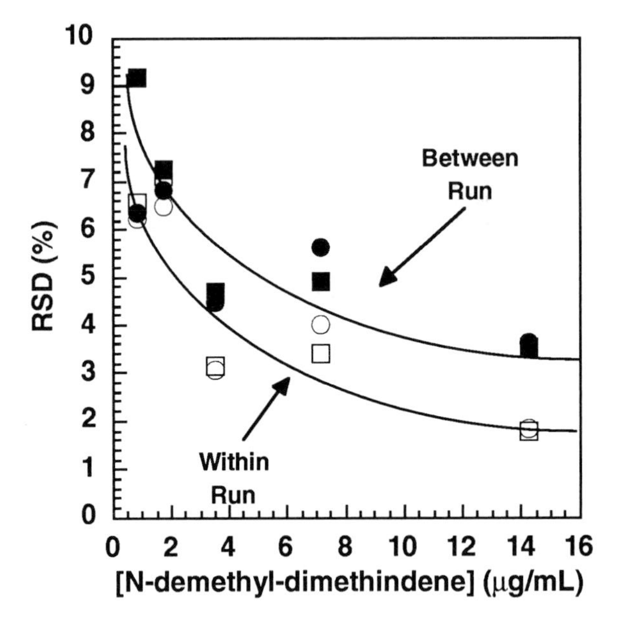


**Figure 2.4**
Within-run and between-run precision for the assay of (R)- (circles) and (S)-N-demethyl-dimethindene (squares) in urine by capillary electrophoresis. Data taken from ref. [8]

Generally the value of the within-run RSD of a method is less than the value of between-run RSD. For example, Fig. 2.4 shows the relationship between the between-run and within-run RSDs for the analysis of N-demethyl-dimethindene in urine by capillary electrophoresis [8]. Figure 2.4 also shows an important feature of chromatographic and related methods of analysis: the precision generally decreases with decreasing analyte concentration, reaching unacceptable levels as the measured signal approaches the noise inherent in the system. However, the precision of a method does not always decrease with increasing concentration. For example, the highest precision of receptor binding assays is generally obtained at intermediate concentrations and decreases at higher and at lower concentrations (e.g. Fig. 2.5).

The fact that the between-run precision of an assay is generally not as good as the within-run precision, at a given concentration, arises from the increased number of steps needed to run an assay on consecutive days compared with the number of steps needed for a single day (e.g. daily preparation of standards, reagents etc.).

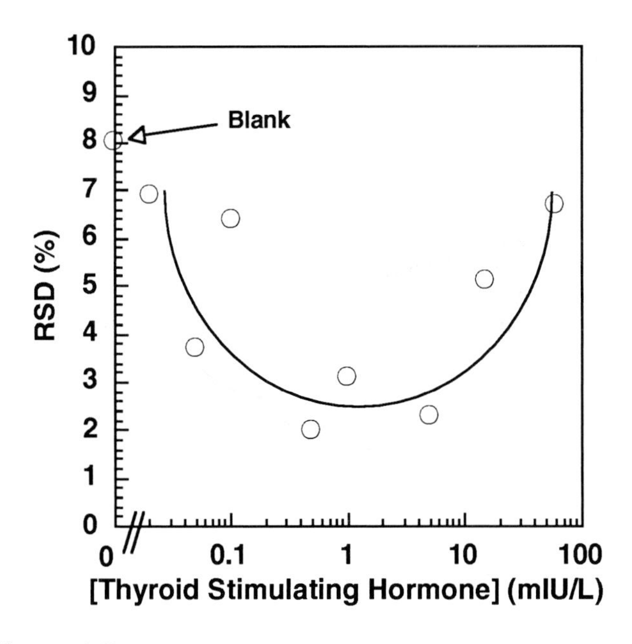


**Figure 2.5**
Within-run precision of the assay of thyroid-simulating hormone by electrochemical enzyme immunoassay. Data taken from ref. [9]

Some analysts have chosen to use the RSD of the slope of the daily calibration curve (see Sec. 2.2.6) as a measure of the between-day precision of an analytical method. This is inappropriate because the slope of the calibration curve is the parameter used to correct for the day-to-day variation in response factor and its RSD only provides an indirect indication of the reproducibility of the method. On the other hand, the day-to-day variability of the response factor can be taken as a measure of the ruggedness of the method (see Sec. 2.2.7).

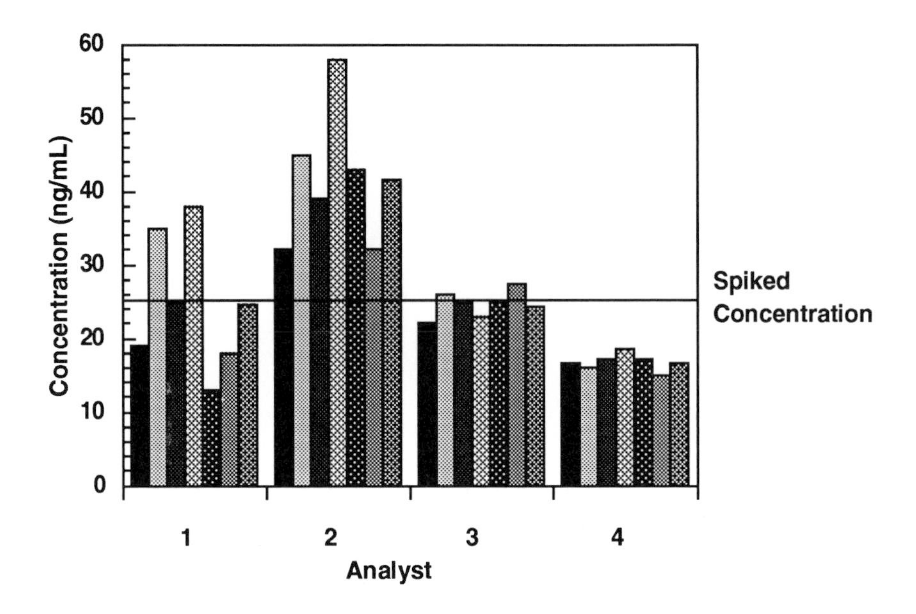


**Figure 2.6**
Results of six replicate assays conducted by four different analysts on urine spiked with a drug at a concentration of 25 ng/mL. This figure illustrates the four possible results: (1) accurate but imprecise, (2) inaccurate and imprecise, (3) accurate and precise, (4) inaccurate but precise

In addition to measuring the within-day and between-day variability, the concepts of accuracy and precision may be used to define the ruggedness or robustness of a method (see also Sec. 2.7). For example, Fig. 2.6 shows the hypothetical results of an experiment designed to determine the effects of different analysts on the determination of a drug spiked into urine at a known (true) concentration of 25 ng/mL. Each analyst demonstrates one of the four possible outcomes: analyst 1, accurate but imprecise; analyst 2, inaccurate and imprecise; analyst 3, accurate and precise; and analyst 4, inaccurate but precise.

It is important to note that increasing the number of measurements on the sample does not necessarily decrease the value of the measured standard deviation. However, as the sample size decreases, so does the uncertainty introduced in using s to estimate the true (population) standard deviation of the method, $\sigma$. To allow for this, the confidence limits for sample mean are given by:

$$\bar{x} = \mu \pm t\left(\frac{s}{\sqrt{n}}\right) \tag{2.10}$$

where the term $s/\sqrt{n}$ is defined as the standard error of the mean and the values of t may be obtained from statistical tables.

### 2.2.2 Calibration

An important step in the validation of any analytical method is the establishment of the mathematical relationship between the measured response ($y_i$) and the concentration of the analyte ($C_i=x_i$). Once the mathematical relationship has been established the analytical instrument or method may be calibrated. The calibration procedure will depend upon the type of method, whether the method is instrumental or non-instrumental, the type of sample, the degree of accuracy and precision required and the concentration range of the analyte or analytes of interest. For the purposes of this discussion, the types of response-concentration functions commonly experienced in pharmaceutical and biomedical analysis have been conveniently divided into those that are linear and those that are non-linear.

#### 2.2.2.1 Linear response functions

The most convenient response function (Fig. 2.7) is one in which the measured quantity ($y_i$) is linearly related to the concentration ($C_i=x_i$) according to the eq.

$$y_i = bx_i + a \tag{2.11}$$

The slope (b) and intercept (a) coefficients are given by eqs. 2.12 and 2.13, respectively:

$$b = \frac{\sum_{i=1}^{i=n}\left\{(x_i - \bar{x})(y_i - \bar{y})\right\}}{\sum_{i=1}^{i=n}(x_i - \bar{x})^2} \tag{2.12}$$

$$a = \bar{y} - b\bar{x} \tag{2.13}$$

where $\bar{x}$ and $\bar{y}$ are the means of the measured responses ($y_i$) and the concentrations ($x_i$), respectively. Figure 2.7 shows the results of a hypothetical experiment to establish the relationship between the peak height obtained by a technique such as liquid chromatography and the concentration of analyte injected. This figure also shows that the relationship between the measured response and the analyte concentration obtained by unweighted least-squares linear regression analysis, must pass through the centroid $(\bar{x}, \bar{y})$ and the intercept (b). Most hand-held calculators and simple computer programs, such as CricketGraph® or DeltaGraph®, readily allow the calculation of the slope and intercept as well as the correlation coefficient (r), which is a measure of the goodness of fit of the data (eq. 2.14). A value of +1 for r indicates perfect correlation and a positive value of the slope (b) and a value of -1 for r indicates perfect correlation and a negative value of the slope (b). A value of 0 for r indictaes no linear correlation between x and y.

$$r = \frac{\sum_{i=1}^{i=n}\left\{(x_i - \bar{x})(y_i - \bar{y})\right\}}{\sqrt{\sum_{i=1}^{i=n}\left\{(x_i - \bar{x})^2(y_i - \bar{y})^2\right\}}} \tag{2.14}$$

If necessary, the statistical significance of the correlation coefficient may be determined using a two-tailed t-test.

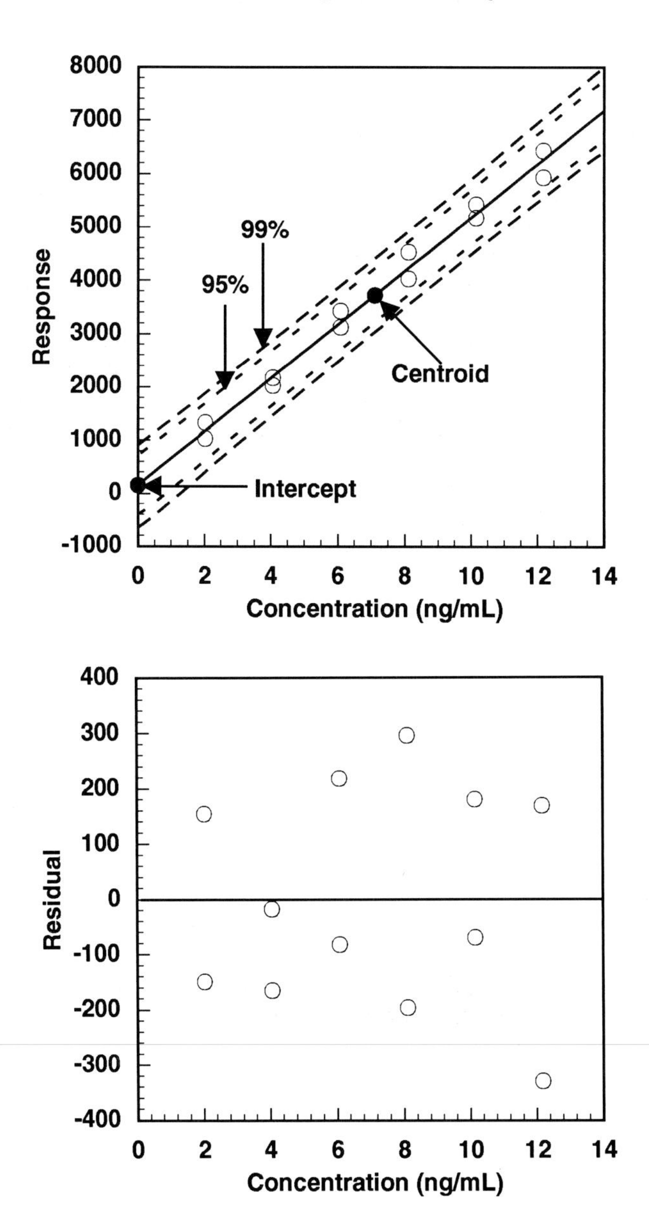


**Figure 2.7**
Regression analysis of representative calibration data showing 95% and 99% confidence intervals (upper) and analysis of residuals (lower)

The value of r obtained for the least-squares linear regression analysis of the sample data in Fig. 2.7 is 0.994. This corresponds to a t-value of 29.05, which is highly significant (P<0.01). Although the value of the correlation coefficient, r, is readily obtained, it should not be relied on to establish the linearity of the calibration data *per se*, because calibration data that have a high degree of curvature can give high values of r (>0.99) and statistically significant values of t. Instead, other approaches should be used to establish the linearity of the response function. The most useful approach to determine whether the calibration relationship is linear is to analyze the residuals ($\delta_{y,x}$, eq. 2.15) or the difference between the observed ($y_i$) and the predicted value of the measured response ($\hat{y}_i$), which should be randomly distributed around a value of zero when plotted against $x_i$ (Fig. 2.7). Any curvature in the data is accentuated by analysis of the residuals.

$$\delta_{y,x} = y_i - \hat{y}_i \qquad (2.15)$$

If the analysis of the residuals (eq. 2.15) suggests curvature in the relationship between response and concentration, a non-linear approach should be considered. In addition to the calculation of residuals, the correlation coefficient (eq. 2.14) and its t-value, a more complete analysis of the data involves the calculation of the standard errors of the slope and intercept, which may be obtained from eqs. 2.16 - 2.18.

$$s_b = \frac{s_{y/x}}{\sqrt{\sum_{i=1}^{i=n}(x_i - \bar{x})^2}} \qquad (2.16)$$

$$s_a = s_{y/x} \cdot \sqrt{\frac{\sum_{i=1}^{i=n} x_i^2}{n\sum_{i=1}^{i=n}(x_i - \bar{x})^2}} \qquad (2.17)$$

where:

$$s_{y/x} = \sqrt{\frac{\sum_{i=1}^{i=n}(y_i - \hat{y})^2}{n - 2}} \qquad (2.18)$$

and $s_b$ is the standard error of the slope (b), $s_a$ is the standard error of the intercept, $\hat{y}$ is the value calculated from the fitted line and $y_i - \hat{y}$ are the residuals. For a perfect linear correlation between x and y, the values of each of the residuals will be 0. The confidence interval for the intercept and the slope may also be calculated (eqs. 2.19 and 2.20, respectively) and their statistical significance determined using a t-test

$$CI(a) = a \pm t \cdot s_a \tag{2.19}$$

$$CI(b) = b \pm t \cdot s_b \tag{2.20}$$

An alternative approach to determining the significance of the y-intercept is to establish whether or not the y-intercept is included in the confidence interval of the calculated values of the response, $\hat{y}_i$ at $x_i$=0 (see Sec. 2.2.2.1.a).

Once the linearity of the response function has been established and the statistical significance of the intercept determined, the decision is made as to whether applications of the method should be based on multiple-point or two-point calibrations. Alternatively, the initial analysis of the response function by least squares unweighted linear regression may indicate that a non-linear response function is more appropriate (see Sec. 2.2.2.2).

### a. Multiple-point calibrations

Multiple point calibration curves are prepared using standard solutions of the analyte in the relevant matrix encompassing the expected concentrations of the analyte in the test sample or samples. The actual range of standard solution concentrations varies with the application and is determined by the range of expected values for the test samples. For the analysis of data by unweighted, least-squares regression the variance of the $y_i$ values is assumed to be approximately equal and the standard concentrations should be evenly spaced (e.g. Fig. 2.7).

The confidence interval of a measured analytical response ($\hat{y}_i$) for a given standard of known concentration, $x_i$, is given by:

$$CI(\hat{y}) = \hat{y} \pm t \cdot s_{\hat{y}} \tag{2.21}$$

where:

$$s_{\hat{y}} = s_{y/x} \cdot \sqrt{\frac{1}{n} + \frac{(x_i - \overline{x})^2}{\sum_{i=1}^{i=n} (x_i - \overline{x})^2}} \quad (2.22)$$

Analysis of calibration data using eqs. 2.21 and 2.22 allows the confidence interval for the measured $y_i$ values in the calibration curve to be determined. However, this does not provide information about the confidence intervals for the concentration of a test sample ($x_{calc}$) determined by comparison with a calibration curve. In this case the confidence interval of $x_{calc}$ is obtained from:

$$CI(x_{calc}) = x_{calc} \pm t \cdot s_{x_{calc}} \quad (2.23)$$

where $s_{x_{calc}}$ is given by the approximation:

$$s_{x_{calc}} = \frac{s_{y/x}}{b} \sqrt{\left\{ \frac{1}{m} + \frac{1}{n} + \frac{(y_{obs} - \overline{y})^2}{b^2 \sum_{i=1}^{i=n} (x_i - \overline{x})^2} \right\}} \quad (2.24)$$

where n is the number of solutions used to prepare the standard curve and m is the number of measurements made on the test sample. Figure 2.7 shows that the confidence interval for a concentration determined from a linear, unweighted calibration curve is narrowest at the centroid $(\overline{x}, \overline{y})$ and increases on either side.

In addition to allowing calculation of the confidence interval for a measured concentration, eq. 2.24 also allows the analyst to calculate the effect of increasing the number of repeat measurements made on the test sample, m (Fig. 2.8). Although Figs. 2.7 and 2.8 show that the standard error of a measurement reaches a minimum at the centroid $(\overline{x}, \overline{y})$, pharmaceutical analysts are usually more interested in the relative error (i.e. the RSD) rather than in the absolute value of the variance. Figure 2.9 shows that the value of $s_{x_{calc}}$ is relatively constant if the value of $x_{calc}$ is equal to or greater than about 25% of $\overline{x}$.

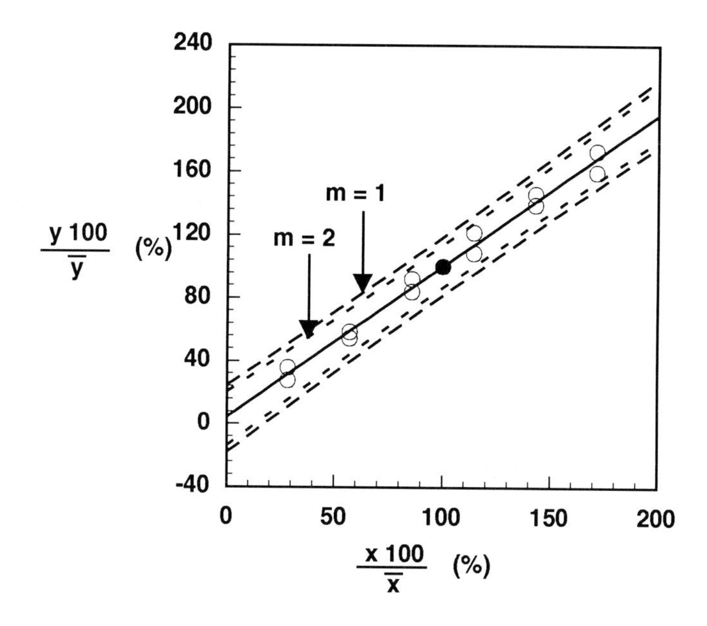


**Figure 2.8**
Effect of number of replicates (m, eq. 2.25) on the 95% confidence interval of a concentration of an analyte calculated from a linear, unweighted calibration curve. The data shown in Fig. 2.7 have been normalized in this figure by dividing the y values by $\bar{y}$ and the x values by $\bar{x}$ (both the x and y values are expressed as a percentage)

Figure 2.9 also highlights the degree of uncertainty that exists in measurements made by extrapolation to values of $x_{calc}$ that are less than the lowest concentration calibration ($x_{min}, y_{min}$) solution. Extrapolation to values that exceed the highest point on the calibration curve $x_{max}, y_{max}$ is not to be recommended and should never be attempted if the linearity of the method has not been established at values greater than $x_{max}, y_{max}$. If the concentration of the test sample substantially exceeds the value of $x_{max}$ then it should be diluted and reanalyzed.

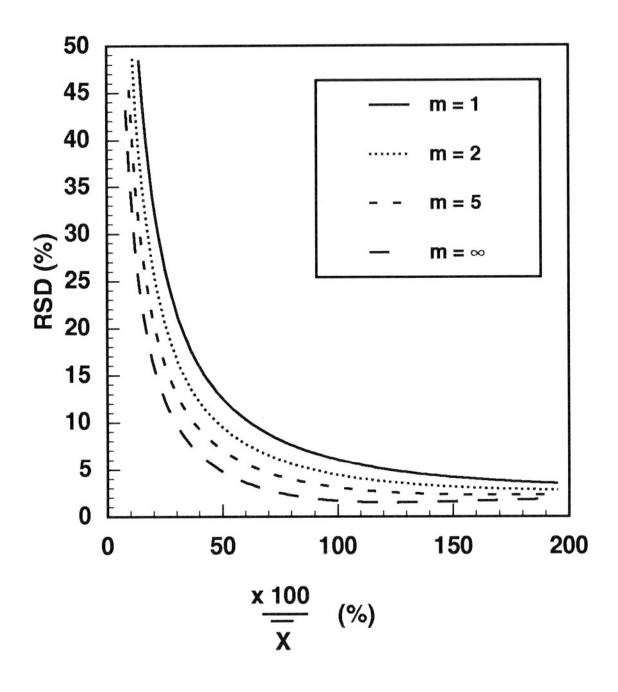


**Figure 2.9**
Effect of number of replicates (m, eq. 2.25) on the RSD of a concentration of an analyte calculated from measured peak height ($y_i$) and a linear, unweighted calibration curve ($x_{calc}$). These plots are based on the data shown in Fig. 2. 7, therefore, the shape of these plots are general but the absolute values are not.

In biological-fluid analysis, the analyst is often faced with very limited samples raising the question as to how much uncertainty is introduced by the analysis of a test sample on the basis of a single measurement. Clearly, eq. 2.24 and Fig. 2.8 show that the confidence interval for a measured concentration will be reduced if the number of measurements made on the test sample is increased. However, the 95% confidence intervals for $x_{calc}$ at the centroid ($x_{calc}/\bar{x}$=100%) for m=1, 2, 5 or ∞ are 5.9, 4.4, 3.7 and 1.7%, respectively and only modest gains in precision are realized if the number of analytical measurements on the test sample is increased from one to two or from two to five. Therefore, if a large number of test samples are to be measured, a substantial amount of work can be saved at only a small cost to the precision if single rather than multiple determinations are made. In contrast very little time will saved if the number of calibration solutions is reduced from a typical value of 12 shown in Fig. 2.7. Furthermore, using a small number of calibration solutions increases the chances of losing the highest or lowest calibration solutions, which define the linear range of the assay. On the other hand, the effect of increasing the number of calibration solutions to a number greater than 12 will have a minimal effect on the confidence interval of the concentrations measured. Equation 2.24 indicates that there is no particular advantage in terms of the precision if all the calibration solutions are prepared at different concentration levels, compared with the preparation of duplicate solutions at fewer concentrations. However, the preparation of duplicate solutions at least at the upper and lower concentration levels eliminates the possibility of losing one of these important calibration solutions due to contamination or spillage (K. Selinger, Glaxo Welcome, personal communication, 1994).

**b. Two-point calibration approaches**

Having established the linearity of a method and the significance of the intercept, the analyst may choose to assay test samples by comparison with a two-point calibration curve[2]. This is the preferred method for the analysis of pharmaceutical samples such as drug substance, raw materials and finished formulations, when the target concentration and the allowable limits are well established. This is also the approach used in most pharmacopeial assays.

In this method of calibration the two standards consist of a blank ($x_1$=0) and a second solution containing the analyte or analytes of interest. Ordinarily the second standard is prepared so that it contains a concentration of the analyte ($x_2$=$C_{std}$) that is as close as possible to the target concentration of the analyte in the test sample. Assays are usually conducted by alternating the measurements of standards and test samples to correct for any instrumental drift that

---

[2] This approach is sometimes referred to as external standardization.

might result in a systematic change in response factor. The response for the standard ($y_{std}$) is then taken as the average of readings obtained for standards measured before and after each sample. This method of calibration is often referred to as "bracketing" of standards. If the instrument drift is small then several samples may measured between the standards. If the intercept is statistically insignificant (i.e. the reading for the blank is close to or effectively zero), which will usually be the case in chromatographic methods of analysis (see Sec. 2.2.3 on specificity and selectivity), then the blank need only be measured once as part of the system suitability test.

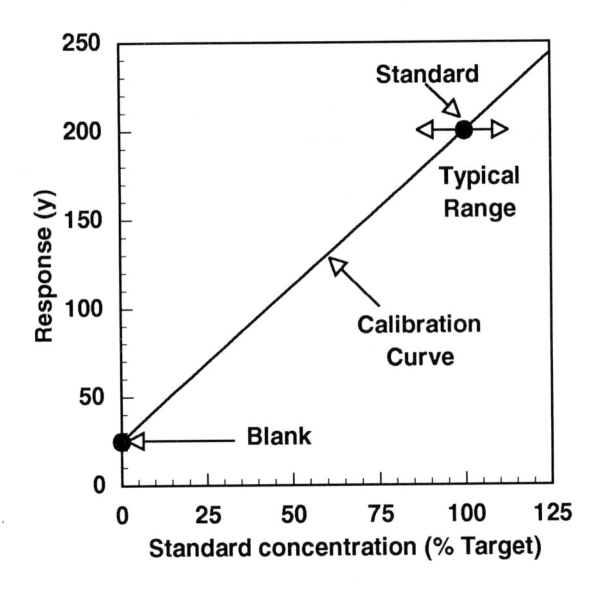


**Figure 2.10**
Two-point calibration curve showing typical range of 90-110% of target concentration

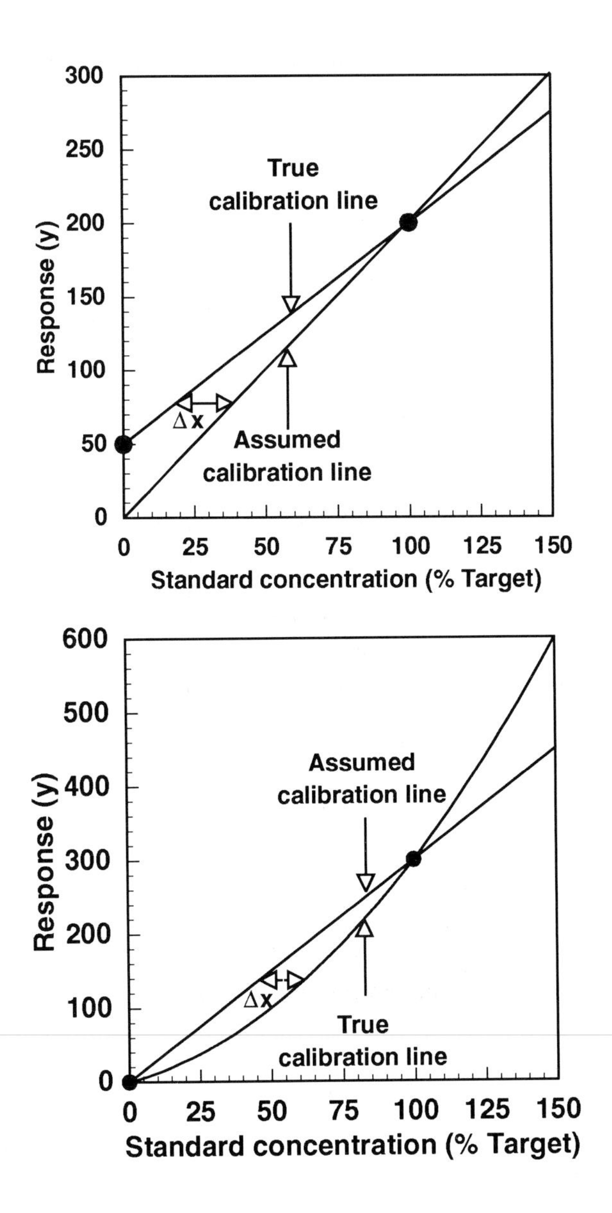


**Figure 2.11**
Potential errors associated with the inappropriate use of a two-point calibration curve. Upper graph: errors due to an intercept; lower graph: errors due to the curvature of the calibration curve

The principle of the two-point calibration approach is demonstrated in Fig. 2.10, for which the equation for the calculation of the concentration of analyte in the test sample ($x_{calc}=C_{calc}$) is given by:

$$C_{calc} = \frac{y_{obs} - \bar{y}_o}{\bar{y}_{std}} \cdot C_{std} \cdot D \cdot \frac{P}{100} \qquad (2.25)$$

where $\bar{y}_0$ and $\bar{y}_{std}$ are the average responses for the blank and the standard bracketing the test sample, respectively; D is the dilution factor and P is the percent purity of the analytical standard. If the intercept is zero, eq. 2.25 reduces to:

$$C_{calc} = \frac{y_{obs}}{\bar{y}_{std}} \cdot C_{std} \cdot D \cdot \frac{P}{100} \qquad (2.26)$$

This method of calibration requires that measurements be made by extrapolation if $y_{obs} > \bar{y}_{std}$. However, this is not cause for concern if the prior validation experiments have demonstrated that the method is linear over the range of the assay. On the other hand, if the actual response function is not linear or a significant intercept is ignored, the potential errors ($\Delta x$) increase with increasing difference between the measured value of the response for the test sample ($y_{obs}$) and the standard ($\bar{y}_{std}$) (Fig. 2.11).

Debesis *et al.* [10] have calculated the maximum method and system precision ($RSD_{max}$) allowable, using the two-point calibration approach for LC assays, as a function of the acceptable assay range (Table 2.1, Fig. 2.12). This was accomplished by assuming that the absolute difference between the true mean and the sample mean is no more than 50% of the specified acceptable range. Thus the method precision is given by:

$$RSD_{max} = \frac{|\bar{x} - \mu|}{z}\sqrt{n} \qquad (2.27)$$

where n is the number of sample measurements and z is taken from tabulated values for normal distribution, i.e. 1.96 or 2.58 for 95 or 99% confidence limits, respectively.

**Table 2.1**
Maximum allowable method RSDs for single or duplicate determinations*

| Acceptance Range (% of claim) | Single Determinations | | Duplicate Determinations | |
|---|---|---|---|---|
| | 95% CI | 99% CI | 95% CI | 99% CI |
| 98.5-101.5 | 0.77 | 0.58 | 1.12 | 0.82 |
| 97-103 | 1.53 | 1.16 | 2.23 | 1.64 |
| 95-105 | 2.55 | 1.94 | 3.72 | 2.74 |
| 90-110 | 5.10 | 3.88 | 7.44 | 5.48 |
| 85-115 | 7.65 | 5.81 | 11.1 | 8.22 |
| 75-125 | 12.8 | 9.69 | 18.6 | 13.7 |
| 50-150 | 25.5 | 19.4 | 37.2 | 27.4 |

*From ref. [10]

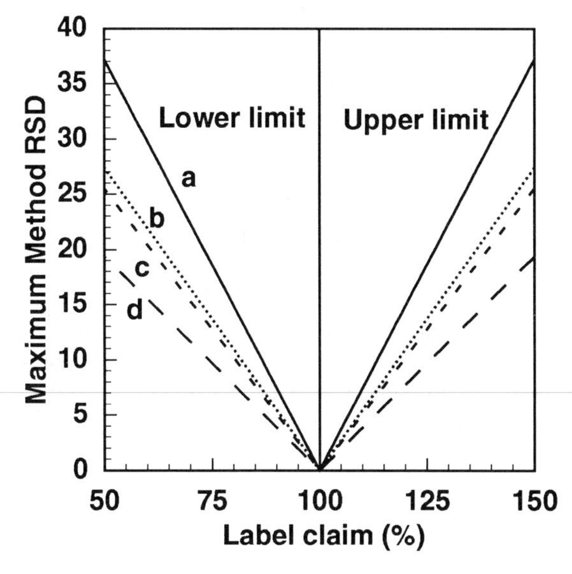


**Figure 2.12**
Relationship between the maximum allowable method precision and acceptable assay limits for single and duplicate determinations. The four lines represent a) CI = 95%, n = 2; b) CI = 99%, n = 2; c) CI = 95%, n = 1; d) CI = 99%, n = 1

Equation 2.27 is particularly useful in deciding whether single or duplicate determinations are necessary. For example, if the acceptance range for an assay is 98.5-101.5% of labeled claim, then the values of $RSD_{max}$ are 0.77% (n=1) and 1.08% (n=2) at the 95% confidence level. At the 99% level, the values are 0.58% (n=1) and 0.82% (n=2). Table 2.1 and Fig. 2.12 show the relationship between the maximum allowable method precision and the acceptable assay ranges for single and duplicate determinations.

### c. Area Normalization

In the absence of authenticated reference standards or where the availability of such standards is very limited, the concentration of related substances in a mixture ($C_{calc,1}$) may be determined by peak area normalization:

$$C_{calc,1} = 100 \frac{\frac{P_{a,1}}{b_1}}{\frac{P_{a,1}}{b_1} + \frac{P_{a,2}}{b_2} + \frac{P_{a,3}}{b_3} \cdots\cdots\cdots} \qquad (2.28)$$

where $P_a$ is the peak area of the individual components (1, 2, 3 ......) in the sample and b is the response function for each component. If authenticated standards are not available, then the assumption must be made that the response function is the same for each component.

Area normalization may be applied to all forms of chromatographic analysis and is particularly suitable for techniques such as gas chromatography with a detector whose response function is independent of the analyte, i.e. b (eq. 2.28) is a constant. This approach is also appropriate for the analysis of data obtained by capillary electrophoresis; however, the peak areas must be corrected for elution time (eq. 2.29) because the velocity of material passing through the detector is inversely proportional to its elution time [11–13].

$$C_{calc,1} = 100 \frac{\frac{P_{a,1}}{b_1 t_1}}{\frac{P_{a,1}}{t_1 b_1} + \frac{P_{a,2}}{t_2 b_2} + \frac{P_{a,3}}{t_3 b_3} \cdots\cdots\cdots} \qquad (2.29)$$

### 2.2.2.2 Non-linear response functions and weighted regression analysis

#### a. Chemical assays

The subject of non-linearity of calibration curves in chromatographic analysis has been treated extensively in the literature [14–24 and Chapter 10], and a number of alternative mathematical models have been proposed, including:

$$y_i = a \cdot e^{b_1 \cdot x_i} \quad (2.30)$$

which is equivalent to:

$$\ln(y_i) = \ln(a) + b_1 x_i \quad (2.31)$$

as well as various quadratic and complex polynomial versions of eqs. 2.31 and 2.32 such as:

$$y_i = b_1 x_i^2 + b_2 x_1 + a \quad (2.32)$$

$$\ln(y_i) = b_3[\ln(x_i)]^2 + b_4 \ln(x_i) + a \quad (2.33)$$

Despite the availability of several possible mathematical models for assay calibration and the ready access to user-friendly computer programs, Burrows and Watson [18] have stated in a recent article on calibration that "*...to obtain optimum results the regression should not only match the actual shape of the response curve but also include appropriate weighting factors to compensate for the error distribution if it is non-homoscedastic. Non-linear calibration routines are often included as part of the capabilities of* (commercial) *chromatographic data systems but may only be suitable for assays covering a small dynamic range and exhibiting a marked degree of curvature. As these regressions are often only available in their unweighted forms they usually impart no improvement in assays covering wide dynamic ranges and showing only small deviations from linearity*" [18]. In addition to the lack of weighted forms of complex non-linear calibration curves, the solutions to the equations for the calculation of concentrations are complex or non-existent. For example, the solution for $x_i$ to the quadratic calibration curve, eq. 2.32, has two roots:

$$x_i = \frac{-b_2 \pm \sqrt{b_2^2 - 4b_1(a - y_i)}}{2b_1} \tag{2.34}$$

Similarly, Burrows and Watson have proposed the complex non-linear eq. 2.35 for calibration of methods using gas chromatography with electron-capture detection (ECD), in which the sensitivity of the detector increases with increasing concentration:

$$y_i = b_5 x_i \ln(x_i) + b_6 x_i + a \tag{2.35}$$

for which no exact solution in $x_i$ exists and concentrations must be read from the calibration graph itself or by calculation using the Newton-Raphson iteration procedure [18].

Calibration of assays over large concentration ranges is often necessary for the analysis of drugs and related substances in biological samples (see also Chapter 10). In this case, the variance may not be homoscedastic (homogeneous) over the entire concentration range and weighted regression analysis on the calibration data is appropriate [18, 25], in which additional weight is given to the lower concentration values. One of the simplest methods of weighting calibration data is according to the reciprocal of the concentration, in which case eq. 2.12 becomes

$$\frac{y_i}{x_i} = b + \frac{a}{x_i} \tag{2.36}$$

The values of the coefficients, a and b, are then obtained by regression analysis of $y_i/x_i$ on $1/x_i$. Weighted regression analysis for the data shown in Fig. 2.7 gives values of 144.9 and 502.0 for a and b, respectively, which compares with values of 156.0 and 501.1 obtained by unweighted analysis of the data. Clearly these changes in the slope and the intercept coefficients have a greater effect on the concentrations calculated ($x_{calc}$) at the lower end of the curve. However the main value of weighted regression analysis is that it narrows the confidence interval for the calculated values of concentration ($x_{calc}$) at the lower end of the curve (Fig. 2.13). The most appropriate weighing factor is generally determined by computer analysis of calibration data (e.g. by SAS® or RS-1), which is also used for the estimation of the standard deviations and the confidence intervals of $x_{calc}$. Figure 2.14 shows the reduction in the RSD for low values of $x_{calc}$ obtained by weighting the calibration data [16]. If weighted regression analysis is used, then the concentrations of the calibration solutions should be unevenly spaced (e.g. if the data are weighted by a $1/x_i$, then the concentrations should be spaced so that intervals for the values of $1/x_i$ are approximately equal, i.e. 1, 2, 5, 10, 20, 50 .....etc.).

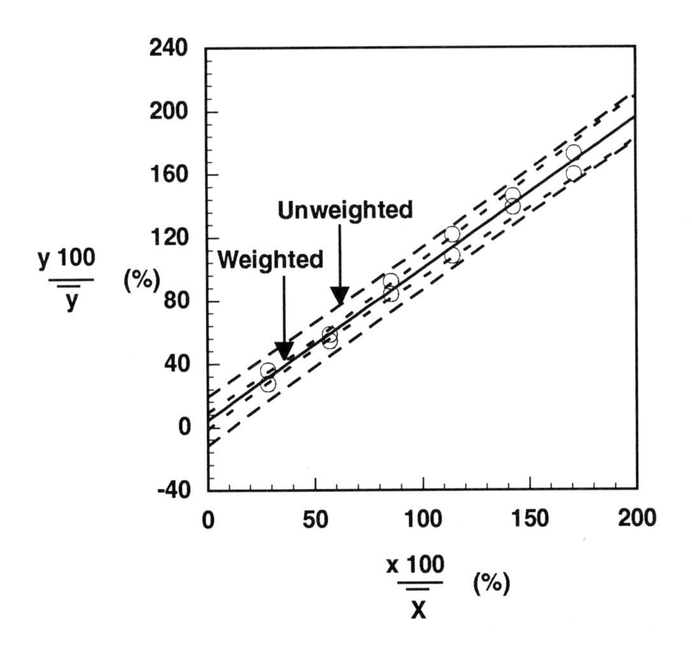


**Figure 2.13**
Comparison of the 95% confidence interval for calculated values of concentration ($x_{calc}$) for weighted and unweighted regression analysis. The data in Fig. 2.7 have been normalized by dividing the y values by $\bar{y}$ and the x values by $\bar{x}$

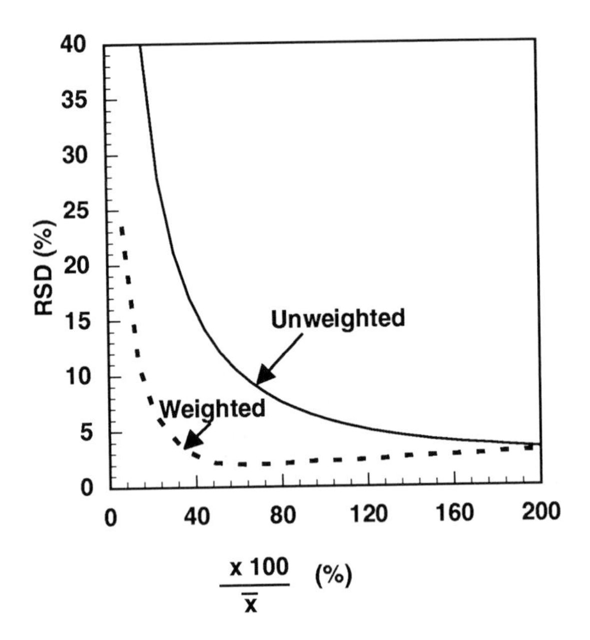


**Figure 2.14**
Comparison of the RSDs of a concentration of an analyte calculated from measured peak height ($y_i$) and a linear calibration curve ($x_{calc}$) for weighted and unweighted regression analysis

## b. Receptor binding assays

Receptor-binding assays, such as radio-immunoassays, are based on the competition between an unlabeled analyte and a labeled ligand for a specific receptor, which may be described by eqs. 2.37 and 2.38:

$$L^* + R \overset{K_d^*}{\rightleftharpoons} B_s^* \tag{2.37}$$

$$L + R \underset{}{\overset{K_d}{\rightleftharpoons}} B_s \tag{2.38}$$

where L is the free ligand, B is the bound ligand, R is the receptor, the sub-script, s, refers to specifically bound ligand and the asterix (*) refers to labeled material [26].

The concentration of bound labeled ligand, [B*], is given by:

$$\begin{aligned}[B^*] &= [B_s^*] + [B_n^*] \\ &= \frac{[R]_0[L^*]}{K_n^*\left(1 + \frac{[L]}{K_d} + [L^*]\right)} + K_n^*[L^*]\end{aligned} \tag{2.39}$$

where $K_n^*$ is the association constant for the non-specific binding of labeled ligand ($B_n^*$), and $R_o$ is the total receptor concentration. For the purposes of calibration, eq. 2.39 may be simplified to give eq. 2.40, which relates the concentration of bound labeled ligand to the concentration of unlabeled analyte added (L):

$$[B^*] = \frac{[B^*]_0}{\left(1 + \frac{[L]}{IC_{50}}\right)} + K^*_n \tag{2.40}$$

where $IC_{50}$ is the concentration of analyte that causes a 50% decrease in the concentration of bound labeled ligand and $[B^*]_o$ is the concentration of bound labeled ligand in the absence of unlabeled ligand. In the absence of non-specific binding the bound ($f_{b^*}$) and free fractions ($f_{f^*}$) of labeled ligand are given by eqs. 2.41 and 2.42, respectively:

$$f_{b^*} = \frac{[B^*]}{[B^*]_o} = \frac{IC_{50}}{(IC_{50} + [L])} \tag{2.41}$$

$$f_{f^*} = \frac{[B^*] - [B^*]_o}{[B^*]_o} = \frac{[L]}{(IC_{50} + [L])} \tag{2.42}$$

Calibration data are usually plotted in the form of the fraction of bound labeled ligand versus log [L] (Fig. 2.15).

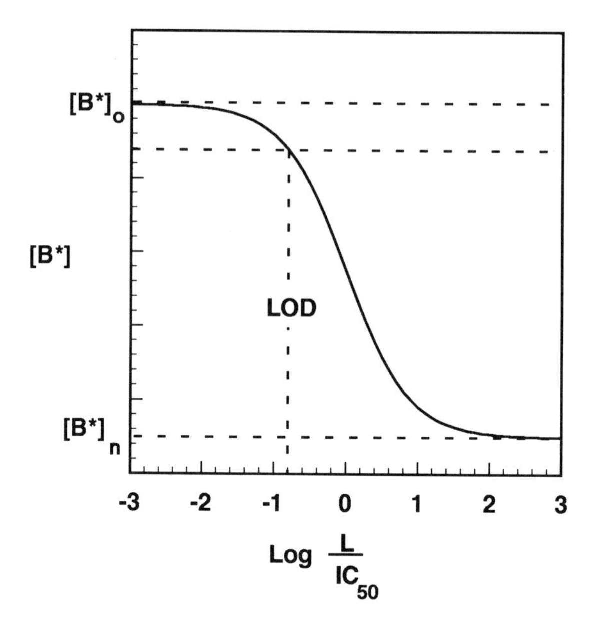


**Figure 2.15**
Calibration curve for a receptor binding assay. Adapted from [26] and reproduced by permission from Elsevier Science Ltd.

### 2.2.2.3 Internal standards

Internal standards are compounds added, in equal concentrations, to all standards and test samples. The original application of internal standards can be traced to gas chromatography [27] where they were used primarily to correct errors arising from manual injection. The practice of including internal standards was later extended to liquid chromatography, mass spectrometry, capillary electrophoresis and other related techniques. When internal standards are included, the analytical response is defined as the peak height (or area) ratio, $P_{r,i}$:

$$P_{r,i} = \frac{y_i}{y_{is}} \tag{2.43}$$

where $y_i$, and $y_{is}$ are the responses obtained for the analyte and internal standard, respectively. Ideally, the internal standard should have physico-chemical and analytical properties that are similar to those of the analyte. They can be used in methods employing calibration curves or two-point calibration approaches. For example if a two-point calibration approach is used with an internal standard eq. 2.26 becomes:

$$C_{calc} = \frac{\overline{P}_{r,obs}}{\overline{P}_{r,std}} \cdot C_{std} \cdot D \cdot \frac{P}{100} \quad (2.44)$$

Because the internal standard is added at the same concentration to all the standards and test samples, its actual concentration is not needed for the calculation of unknowns. In principle, the analytical response of the internal standard should be linearly related to concentration. However, any errors arising from non-linearity are likely to be small unless the recovery of the analytes and the internal standard varies widely between samples.

Whereas the original intention was to correct for variation in injection volumes in gas chromatography, the present-day automatic injectors used in gas chromtaography and liquid chromatography make the use of internal standards for many applications unnecessary. Moreover, the additional steps involved in the volumetric addition of an internal standard may actually reduce the precision of the method (see Sec. 2.2.1). However, the use of internal standards in capillary electrophoresis is strongly recommended because the reproducibility of automated injectors for capillary electrophoresis does not yet approach that achievable in gas chromatography or liquid chromatography [28].

Applications where internal standards are essential include assays involving sample preparation where the extraction efficiencies are low or those involving chemical derivatizations with low yields of reaction. On the other hand if the sample preparation step is straightforward and the extraction efficiency is close to 100% (e.g. simple protein precipitation for the pre-treatment plasma samples) then an internal standard is advisable but, in certain circumstances, may be unnecessary. The appropriateness of adding an internal standard should be determined as part of the method validation.

### 2.2.3 Selectivity and specificity[3]

Selectivity describes the ability of analytical method to differentiate various substances in the sample and is applicable to methods in which two or more components are separated and quantitated in a complex matrix. Thus the term selectivity is appropriately applied to chromatographic techniques in which the components of a mixture are physically separated from each other. Selectivity may also be used to describe spectroscopic or spectrophotometric methods in which separate signals are obtained for the different components in a mixture. In contrast to selectivity, specificity, describes the ability of the method to measure unequivocally the analyte of interest in the presence of all other components, which may be expected to be present. Thus, the term specificity is appropriately applied to analytical techniques in which only a single parameter can be measured: examples include the measurement of radioactivity in a radioimmunoassay or the volume of titrant in a titration. The selectivity or the specificity of a method is compromised by the presence of potential interferences including related compounds (degradants, metabolites, impurities etc.) and components of the matrix (formulation excipients, endogenous substances etc.).

---

[3] The descriptions given here represent the International Union of Pure and Applied Chemistry (IUPAC) definitions of specificity and selectivity described in the official journal of the Union (*Pure Appl. Chem.*, **35**, 553-556, (1983)) as follows: *"It is therefore proposed to use the adjectives selective and specific and the substantives selectivity and specificity, when not followed by other substantives, as means to express, qualitatively the extent to which other substances interfere with the determination of a substance according to a given procedure. In this connection specificity is considered to be the ultimate of selectivity, it means that no interferences are supposed to occur".* Notwithstanding the IUPAC defintions of specificity and selectivity, these terms are often used interchangeably. For example the USP definition of specificity is "*....its ability to measure accurately and specifically the analyte in the presence of components that might be expected to be present*" and the recent ICH Guidelines define specificity as *"...the ability to assess unequivocally the analyte in the presence of components which may be expected to be present"* - definitions more akin to the classical definition of selectivity.

### 2.2.3.1 Chromatographic methods

The selectivity of a chromatographic method may be defined by the use of relative retention indices of which the most widely used is the selectivity factor, $\alpha$:

$$\alpha = \frac{k'_2}{k'_2} \tag{2.45}$$

where the capacity ratio of a given peak $k'_i$, is related to its retention time, $t_i$, and the time for an unretained compound to elute, $t_o$, by:

$$k_i = \frac{t_i}{t_o} - 1 \tag{2.46}$$

Substitution of eq. 2.45 into eq. 2.46 gives:

$$\alpha = \frac{t_2 - t_o}{t_1 - t_o} \tag{2.47}$$

Chromatographic resolution is also affected by column efficiency, which is usually defined by the number of theoretical plates (N):

$$N = \left(\frac{t_i}{\sigma_i}\right)^2 \tag{2.48}$$

Since the width of a Gaussian peak at its base ($w_i$) corresponds to approximately $4\sigma$ (Sec. 2.1.3, Fig. 2.3), the number of theoretical plates may be calculated from:

$$N = 16\left(\frac{t_i}{w_i}\right)^2 \tag{2.49}$$

Measurement of $w_i$ according to eq. 2.49 requires extrapolation of tangents drawn at the inflection points on the trailing and leading edges of peak to the baseline (Fig. 2.16), which is subject to error. This has led to an alternative form of eq. 2.49, which relies upon measurement of the peak width at half height, $w_{0.5}$, (eq. 2.50).

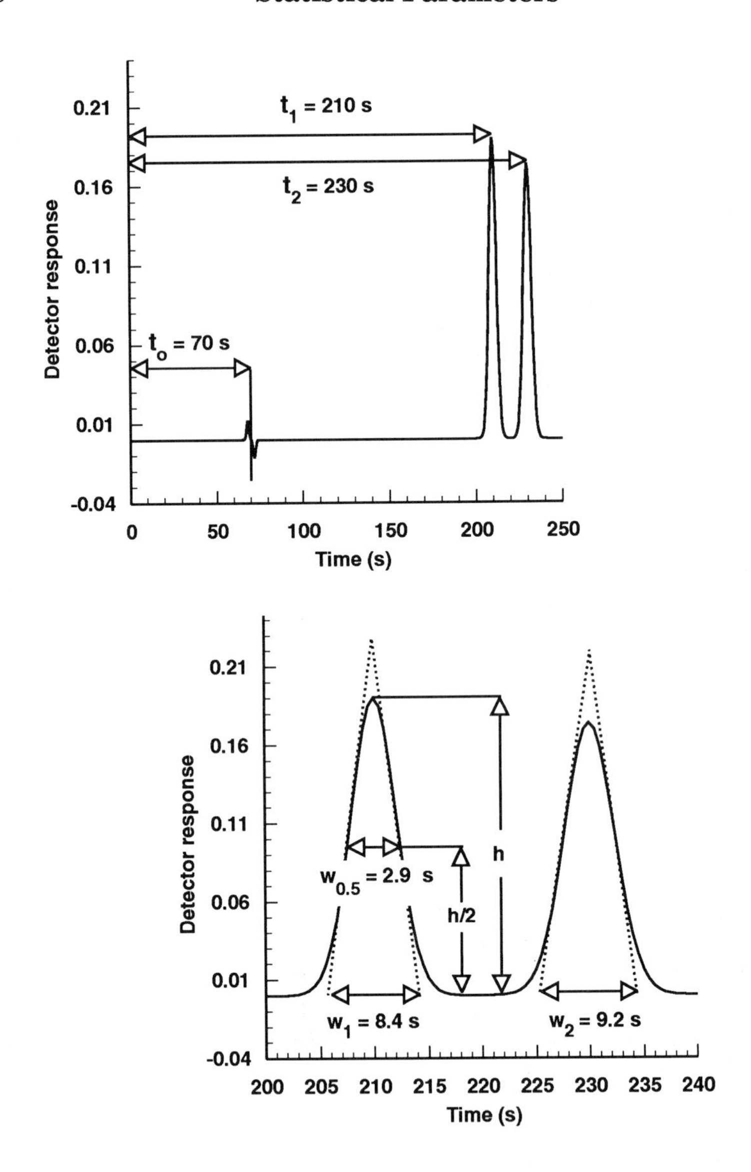


**Figure 2.16**
Parameters for the calculation of chromatographic resolution

$$N = 5.54\left(\frac{t_i}{w_{0.5}}\right)^2 \quad (2.50)$$

For Gaussian peaks, eqs. 2.49 and 2.50 are equivalent. However, for peaks that are significantly asymmetric, then eq. 2.50 tends to underestimate the true column efficiency, because peak tailing and fronting affect the base of the peak to a greater extent than the center. Equations 2.49 and 2.50 are the definitions of column efficiency preferred by the USP. The BP [29] and the EP [30, 31] define column efficiency in terms of the number of theoretical plates per unit column length (plates/meter), $n$:

$$n = \frac{N}{L} = \frac{5.54}{L}\left(\frac{t_i}{w_{0.5}}\right)^2 \quad (2.51)$$

The resolution of two components in a mixture (Figure 2.16) is defined in the USP [32] and in the BP [29] and EP [30, 31] by eqs. 2.52 and 2.53, respectively:

$$R_s = 2\frac{(t_2 - t_1)}{w_2 + w_1} \quad (2.52)$$

$$R_s = 1.18\frac{(t_2 - t_1)}{w_{0.5,2} + w_{0.5,1}} \quad (2.53)$$

The differences in these two equations arise from the different ways in which the widths of the peaks are measured, the USP preferring the width at the base and the BP and the EP preferring the width at half-height. As with the measurement of column efficiency, the two methods for calculations of resolution are equivalent if the peaks are Gaussian, but the USP approach is (eq. 2.52) is less susceptible to errors arising from peak asymmetry than are the BP or EP approaches (eq. 2.53). Errors arising from the manual measurement of column efficiency and resolution may be reduced by the automatic calculation of these parameters using commercial chromatographic software; however, the analyst should be familiar with the algorithms used because different software packages can give different results. This is a particular problem when transferring chromatographic methods between laboratories.

Figure 2.17 shows the effect of resolution on the detectability of a secondary component eluting close to a major component. If the resolution is equal to or greater than 2, then the trace component is detectable at all concentrations. However, if the separation deteriorates only slightly ($R_S<2$) then the detectability of the secondary component becomes a function of its concentration. As the resolution approaches a value of one, impurities present at 0.5% or less are indistinguishable from the peak tail. Clearly, the problems associated with the detection of trace impurities shown in Fig. 2.17 are accentuated by any tailing of the major component. Consequently, the preferred elution order in trace analysis in LC is one in which the minor component elutes before the major component.

The use of the peak separation index, S, (Fig. 2.18 and eq. 2.54) represents a useful alternative to resolution for the description of chromatographic or electrophorectic separations. This empirical approach is particularly useful for situations where one or both of the peaks are asymmetric and has been widely used for the optimization of chiral separations and trace analysis. The peak separation index approach is based on the relative depth of the valley between two adjacent peaks:

$$S = \frac{A}{A + B} \cdot 100 \qquad (2.54)$$

Two approaches have been described for the calculation of the parameters, A and B (Fig. 2.18). The first approach, which is preferred for separation of two components present in similar proportions, involves drawing a line perpendicular to the x-axis through the minimum to intersect a tangent drawn between the apices of the two peaks. For trace analysis, drawing a tangent between the apices of the two peaks is difficult, so a line is drawn parallel to the baseline through the apex of the minor peak to intersect a line drawn through the valley perpendicular to the x-axis (Fig. 2.18).

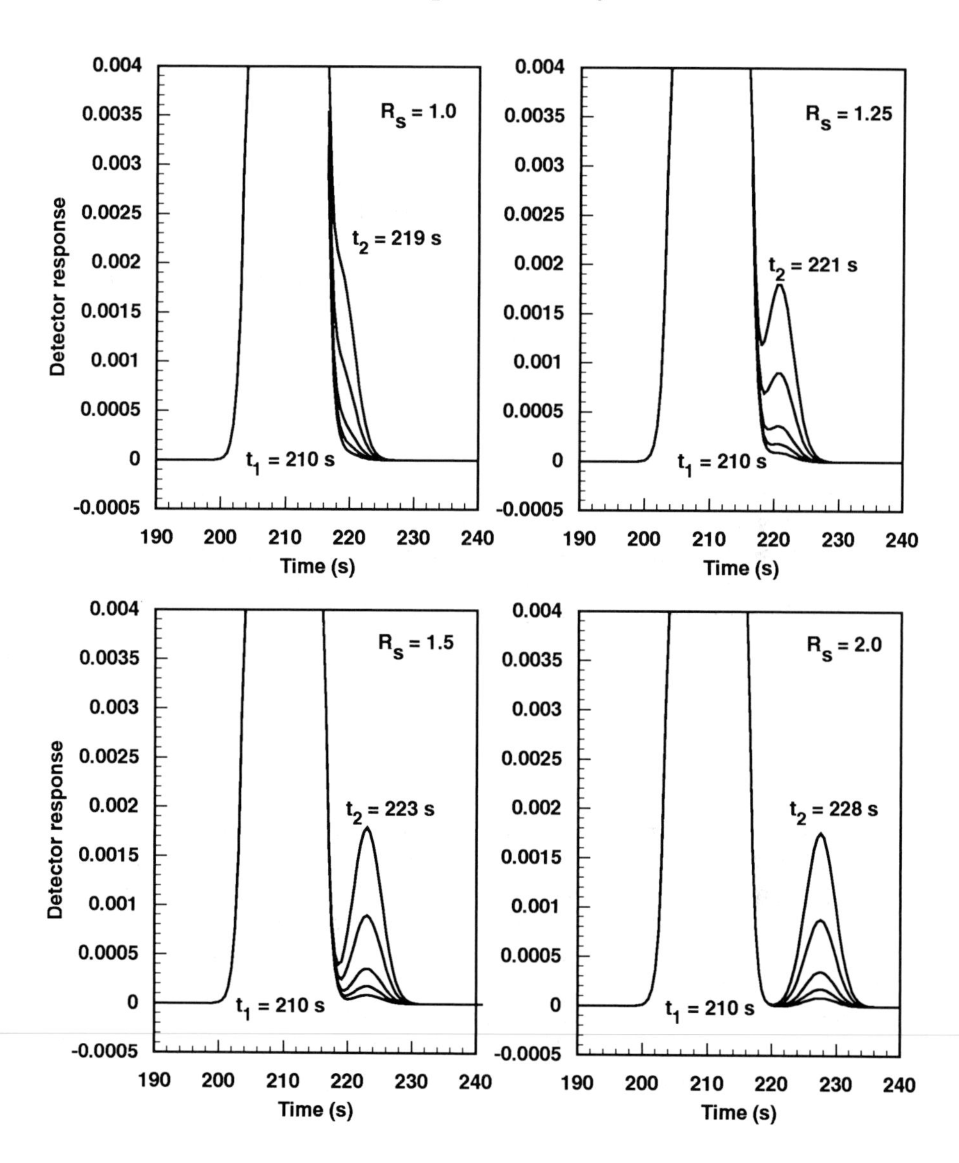


**Figure 2.17**
Effect of impurity levels (0.05, 0.10, 0.20, 0.50 and 1.0%) on chromatographic resolution

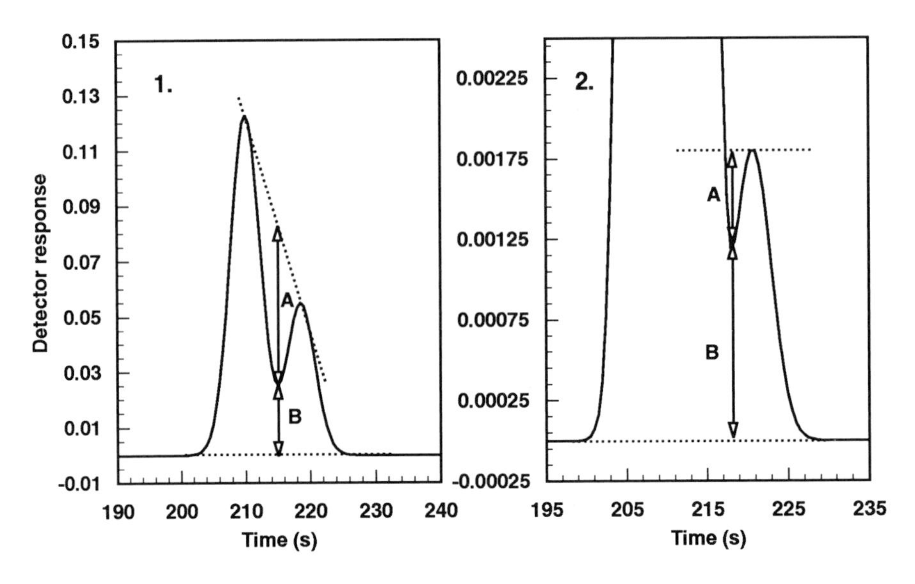


**Figure 2.18**
Peak separation indices. 1. is suitable for separations of two components present in similar proportions and 2. is more suitable for trace components

Although these approaches are useful to assess the quality of a chromatographic separation, excessive tailing of peaks is indicative of poor chromatography, which can lead to unreliable integration of peak areas and poor precision [32]. The USP, BP and EP define the tailing factor, T, as:

$$T = \frac{w_{0.05}}{2 f_{0.05}} \tag{2.55}$$

where $w_{0.05}$ is the width of the peak at 5% of the height and f is measured according to Fig. 2.19. The pharmacopeial definition of the tailing factor varies slightly from the classical definition of the peak asymmetry factor, $A_s$ [33], which is generally measured at 10% of the peak height and is given by:

$$A_s = \frac{g_{0.1}}{f_{0.1}} = \frac{w_{0.1} - f_{0.1}}{f_{0.1}} \tag{2.56}$$

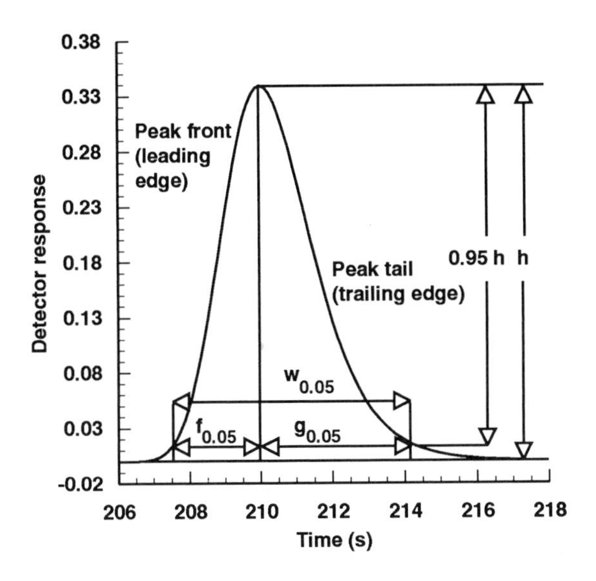


**Figure 2.19**
Parameters for the calculation of peak asymmetry

Both the peak tailing factor, T, and the asymmetry factor, $A_S$, have values of one for perfectly symmetrical peaks, values of greater than one for tailed peaks and less than one for fronted peaks. It should be noted that Foley and Dorsey [33] have recommended the use of eq. 2.57 for the precise calculation of column efficiency, which includes a correction for peak asymmetry:

$$N_{sys} = \frac{41.7\left(\frac{t_i}{w_{0.1,i}}\right)^2}{A_S + 1.25} \tag{2.57}$$

The selectivity of chromatographic methods and other techniques such as mass spectrometry are normally assessed by spiking experiments in which the analyte and all known or suspected potential intereferences are added to the sample matrix (a placebo) at appropriate concentrations. The spiked placebo is then compared with the representative blanks. The elution times of all significant compounds are then recorded and the resolution of the critical pair or pairs of peaks determined. The critical pair or pairs are then used to set the limits for the resolution checks to be used as part of the system suitability tests.

The main problem in the determination of assay selectivity and specificity arises from unknown interferences that are present in the test samples but not in the placebo. Thus the determination of peak homogeneity plays a key role in the assessment of chromatographic selectivity. Whereas the mass spectrometer provides a very high degree of selectivity in gas chromatography, the diode-array detector (DAD) and the use of spectral deconvolution techniques are the principal tools for the determination of peak purity in liquid chromatography. Unfortunately, spectral deconvolution requires that the UV/visible spectra of the co-eluting peaks be different and commercially available systems are generally incapable of detecting any co-eluting interference that is present at concentrations much less than 1%. With the increasing availability of reliable interfaces, on-line mass spectrometric detection is expected to play an increasingly important role in the determination of peak purity in liquid chromatography.

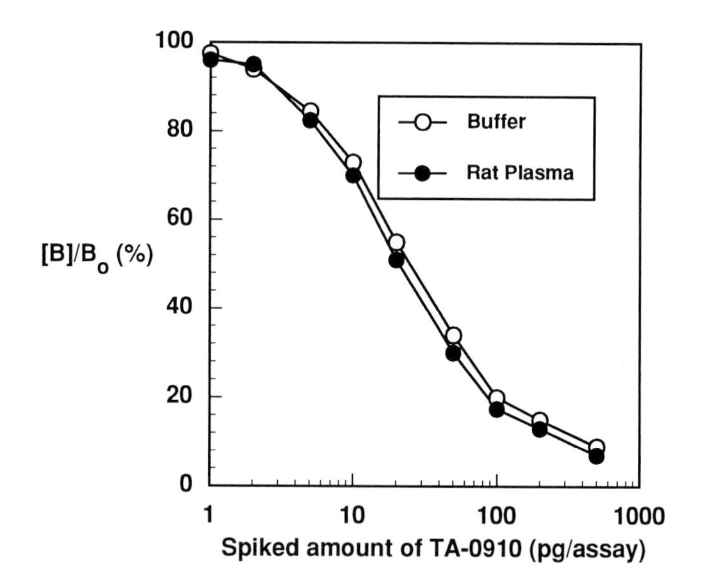


**Figure 2.20**
Binding curves for the radioimmunoassay of a thyroid-releasing hormone analog (TA-01910) in buffer and in plasma. Reproduced from [34] with permission from Elsevier Science

The specificity of immunoassays and other receptor binding assays is determined in the same way as the selectivity of chromatographic methods by spiking experiments and by comparison with the appropriate blanks. For example, Fig. 2.20 shows the receptor-binding curve for the assay of the thyroid-releasing hormone analog, TA-01910 in buffer and rat plasma, showing the absence of interference from the matrix. The problem of matrix interference in receptor binding assays is of particular concern for the analysis of drugs in plasma because it involves plasma-protein binding of both the labeled and the unlabeled ligand [26] (Fig. 2.21). Interferences from related substances (metabolites, degradants etc.) is generally assessed by the comparison of the $IC_{50}$ values of potential interferences with the $IC_{50}$ value for the analyte of interest.

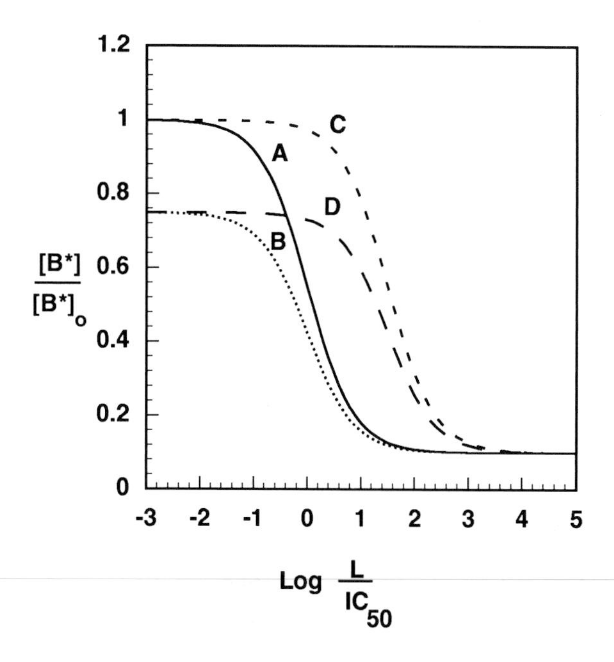


**Figure 2.21**
Effect of plasma-protein binding on the binding of labeled and unlabeled ligand to the receptor in an immunoassay. A: control; B: protein binding of labeled ligand; C: protein binding of unlabeled ligand; D: protein binding of labeled and unlabeled ligand. Adapted from [26]

### 2.2.4 Limit of detection (LOD)

The LOD of an analytical method is an important parameter if quantitative measurements are to be made at concentrations close to it. It is therefore essential to define the LOD of methods for the trace analysis of substances in pharmaceutical and biomedical samples, for example trace impurities in a bulk drug sample, degradation products in a finished pharmaceutical product or a drug and its metabolites in a biological fluid. Highly precise measurements are impossible at concentrations close to the LOD (see secs. 2.2.4.1-2.2.4.3). The LOD is a relatively unimportant parameter for methods, where by necessity the degree of precision is very high, for example methods for the quality control of the active substance in the bulk drug form or in a finished pharmaceutical formulation, because such methods will be used at concentrations substantially greater than the LOD. Accordingly, the United States Pharmacopeia defines the LOD as "*.... a parameter of limit tests. It is the lowest concentration of analyte that can be detected, but necessarily not quantitated, under the stated experimental conditions. Thus limit tests merely substantiate that the analyte concentration is above or below a certain level....*" [35].

#### 2.2.4.1 Spectroscopic methods

The limit of detection (LOD) of an analytical method may be defined as the concentration which gives rise to an instrument signal that is significantly different from the blank [36]. The usual convention defines the LOD as the concentration giving rise to a signal that is three times the standard deviation of the blank (i.e. a signal:noise (S:N) ratio of 3:1). For spectroscopic techniques or other methods that rely upon a calibration curve for quantitative measurements, the IUPAC approach employs the standard deviation of the intercept ($s_a$), which may be related to the LOD and the slope of the calibration curve, b, by:

$$\mathrm{LOD} = \frac{3 \cdot s_a}{b} \tag{2.58}$$

### 2.2.4.2 Chromatographic methods and related techniques

The standard deviation of the blank, $s_a$ may be considered as a measure of the noise (N) in the system. For instruments that produce a continuous, electronic output, the noise may be measured from the peak-to-peak variation ($N_{p\leftrightarrow p}$)in the baseline signal:

$$s_a = N_{p\leftrightarrow p} \tag{2.59}$$

Substituting eq. 2.59 into 2.58 gives:

$$LOD = \frac{3 \cdot N_{p\leftrightarrow p}}{b} \tag{2.60}$$

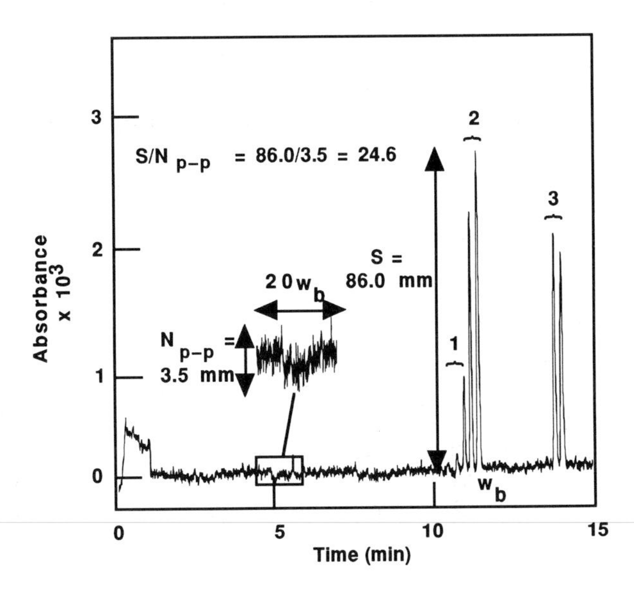


**Figure 2.22**
Sample calculation of LOD for capillary electrophoresis using the separation of N-demethyl-dimethindene (1), dimethindene (2) and 6-methoxydimethindene (3) from human urine. Electropherogram reproduced from [8] with permission from Elsevier Science

The British Pharmacopeia has adopted the recommendation of Foley and Dorsey [20] that the noise ($N_{p\leftrightarrow p}$) be measured over a representative section of baseline equal to 20 times the width of the analyte peak ($w_i$). In the example shown in Fig. 2.22, the signal-to-noise ratio ($S{:}N_{p\leftrightarrow p}$) for the R-isomer of methindene separated from urine by capillary electrophoresis is 24.6:1 [8]. The peak measured in Fig. 2.22 represented approximately 10 ng/mL in the urine specimen. Therefore the LOD for R-methindene in urine was approximately 1 ng/mL assuming that the response was linear with respect to concentration between the 1 ng/mL and the LOD.

#### 2.2.4.3 Receptor binding assays

Šmisterova *et al.* [26] have shown that the LOD of receptor binding assays, such as immunoassays may be calculated using a t-test to determine the smallest concentration of analyte at which the concentration of labeled ligand bound is statistically different from the concentration of labeled ligand bound in the absence of the analyte. They have shown [26] that the LOD of a receptor-binding assay depends on the concentration of labeled ligand, the value of the binding constant of the ligand, $K_d$, and the error inherent in the measurement of the concentration of bound labeled ligand in the absence of unlabeled ligand (Fig. 2.15).

### 2.2.5 Limit of quantitation (LOQ)

For many pharmaceutical applications the limit of quantitation (LOQ) is generally a more useful parameter than the LOD. The LOQ is the lowest concentration in a sample that may be measured with an acceptable level of accuracy and precision. As with the LOD, the LOQ is relevant only in trace analytical methods when measurements are being made at concentrations close to that limit. The United States Pharmacopeia defines the LOQ as *"....a parameter of quantitative assays for low levels of compounds in sample matrices, such as impurities in bulk drug substances and degradation products in finished pharmaceuticals"* [35]. The empirical definitions of LOQ are arbitrary and vary with the type of method employed and the nature of the sample. They also depend on whether the method is instrumental or non-instrumental.

The IUPAC approach [37] is the most appropriate method for the calculation of the LOQ for a spectrophotometric method. In this case the LOQ is defined as the concentration that gives rise to a signal, which is ten times the standard deviation of the blank, i.e.:

$$LOQ = \frac{10 \cdot N_{p \leftrightarrow p}}{b} \qquad (2.61)$$

It follows, according to IUPAC definitions, that:

$$LOQ = 3.\dot{3} \cdot LOD \qquad (2.62)$$

An alternative method (c.f. eq. 2.61 and 2.62) is generally employed for the analysis of drugs in biological fluids by chromatographic techniques. In those cases the LOQ is defined as the concentration at which the analyte of interest may be measured with an RSD of less than a critical value (e.g. 20% [38]). The subject of precision in biological fluid assays is discussed in more detail in Chapter 10. The LOD and the LOQ should be measured in the appropriate medium because interference from components in the matrix may cause these values to be higher than the values that may be obtained from aqueous standard solutions. However, it is important to note that co-eluting peaks in chromatographic techniques contribute a positive bias and therefore contribute to the systematic error and not to the random error or the precision of the method. Thus a distinction may be made between the LOQ that is determined by the precision and the LOQ that is determined by the accuracy.

### 2.2.6 Sensitivity

In contrast to the limit of detection, assay sensitivity (S) is defined as the ability of the assay to distinguish different concentrations and is given by:

$$S = \frac{dR}{dC} \left( or = \frac{dy_i}{dx_i} \right) \qquad (2.63)$$

For assays where the response function is linear, the sensitivity is constant with respect to concentration and is equal to the slope of the calibration curve (b, eq. 2.12). Substitution of eq. 2.63 into eq. 2.60

gives eq. 2.64, which describes the relationship between sensitivity and the LOD of assay. This relationship holds only for assays with linear response functions; no such relationship exists for assays with non-linear response functions because the sensitivity varies with concentration (Fig, 2.23).

$$S=\frac{3 \cdot N_{p\leftrightarrow p} \cdot s_a}{LOD} \qquad (2.64)$$

In contrast to linear response functions, the sensitivity of the assay changes with analyte concentration when the response function is non-linear. For example, in receptor-binding assays, the sensitivity of the assay is maximum when $[L]=IC_{50}$ and approaches zero as the concentration of labeled ligand bound to the receptor approaches 1 at low analyte concentrations and as it approaches 0 at high analyte concentrations. Fig. 2.23 compares the sensitivities obtained for assays with linear and non-linear response functions.

### 2.2.7 Ruggedness and robustness

There is a certain lack of uniformity and certainly a degree of confusion in the literature and in the various compendia and regulatory guidelines regarding the definitions of ruggedness and robustness [10, 35, 36, 39, 40]. Whereas in some circles the ruggedness and robustness are synonymous, the USP defines ruggedness [35] as *"the degree of reproducibility of test results obtained by the analysis of the same samples under a variety of normal test conditions, such as different laboratories, different analysts, different instruments, different reagent lots, different elapsed assay times, different assay temperatures, different days, etc. Ruggedness is normally expressed as the lack of influence on test results of operational and environmental variables of the analytical method. Ruggedness is a measure of reproducibility of test results under normal, expected operational conditions from laboratory to laboratory and from analyst to analyst"*. In the USP the robustness of an analytical procedure is defined as *"a measure of its capacity to remain unaffected by small but deliberate variations in method parameters and provides an indication of its reliability in normal usage"*.

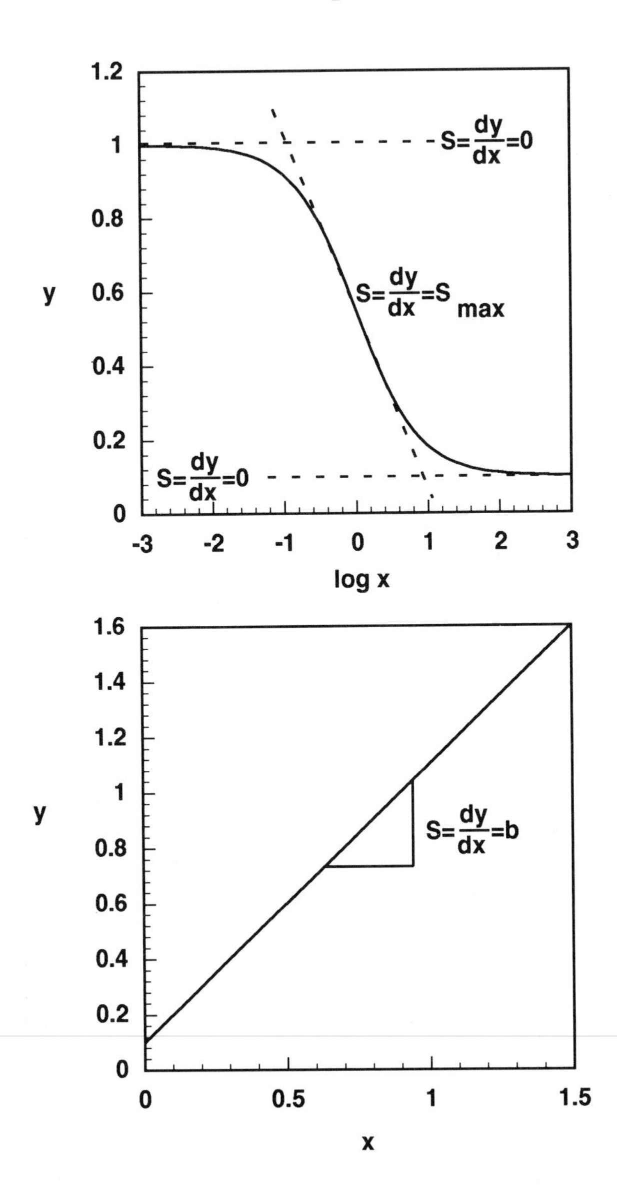


**Figure 2.23**
Sensitivities ($dy_i/dx_i$) for non-linear (upper) and linear (lower) response functions

The USP goes on to indicate that assay ruggedness should be determined by *"analysis of aliquots of homogenous lots in different laboratories, by different analysts, using operational and environmental conditions that may differ but are still within the specified parameters of the assay. The degree of reproducibility of test results is then determined as a function of the assay variables. This reproducibility may be compared to the precision of the assay under normal conditions to obtain a measure of ruggedness of the analytical method"*.

The detailed definitions of ruggedness and robustness described in the USP are in contrast to the Guidelines on Validation of Analytical Procedures for Pharmaceuticals published in 1994 by the International Conference on Harmonization (ICH) [41], which defines the robustness of an assay *"as a measure of its capacity to remain unaffected by small, deliberate variations in method parameters and provides an indication of reliability during normal usage"*. No specific mention is made in the ICH Guidelines to assay ruggedness, which is alluded to in the definitions of precision: *"<u>Repeatability</u> expresses the precision under the same operating conditions over a short period of time. Repeatability is also termed inter-assay precision. <u>Intermediate precision</u> expresses within laboratory variations: different days, different analysts, different equipment etc. <u>Reproducibility</u> expresses the precision between laboratories (collaborative studies)"*.

Notwithstanding the present ambiguity of definitions, it is convenient to apply the term assay ruggedness to the variation or errors in assay results arising from different operation conditions and to apply the term assay robustness to the ease with which the critical parameters of the assay may be reproduced. It also follows that the ruggedness of an assay is influenced by its robustness. For example, an LC assay demonstrating very good precision in one laboratory might yield very poor precision in a second laboratory if the separation obtained by the analyst in the second laboratory gave very tailed peaks. Such examples of the robustness in LC assays are commonplace because of the number of environmental and operational factors that can affect the reproducibility of the desired separation.

Degradation of chromatography columns is a common cause of assay deterioration and lack of robustness [40]. Thus, any chromatographic method should include a description of column washing and storage conditions as well as the inclusion of appropriate system suitability tests (SSTs) for the determination of column performance. In addition, the biggest cause of poor inter-laboratory reproducibility of chromatographic separations is large batch-to-batch variability of chromatography columns. An additional problem is the difference in the chromatographic performance of stationary phase purported to have the same chemical composition [40]. Thus during the method development the influence of different columns, even from the same manufacturer, should be examined and the effects clearly described in any published report.

Other factors that will affect the reproducibility of separations include the effects of mobile phase composition and temperature. The need to document the sensitivity of an assay to small changes in environmental and operational conditions is particularly important if that method is submitted in support of a New Drug Application (NDA). In 1986, Sheinin [40] described examples of problems encountered by scientists at the FDA in attempts to reproduce LC methods submitted in support of various NDAs. These problems included:

- Inadequate column specifications - separation could not be reproduced using column specified

- Variations in the mobile phase - separation was very sensitive to small changes in mobile phase composition (organic modifier)

- Inadequate method development - separation on column described in method was poor, other columns tried by FDA laboratories were more suitable

- Column deterioration - resolution of critical peak pairs degraded within a few hours after initiation of assay

- Lack of samples - authenticated samples of related compounds not supplied

The degree of ruggedness and robustness of a particular analytical method will depend largely on the intended application. For example, the degree of ruggedness and robustness needed for an assay of very limited application to study a new chemical entity in the very early stages of development is much less than would be required to control the quality of a marketed product being manufactured at several different locations throughout the world. In the first case it may only be necessary to examine the effects of changing the chromatography column and making small changes in mobile phase composition on the resolution of the critical pair of peaks in the chromatogram. In the second case an extensive muti-laboratory collaborative study would be required.

The Canadian Guidelines on assay validation [42] define three levels of ruggedness testing depending on the anticipated future application of a new method (Table 2.2). However, it should be noted that the Canadian Guidelines do not distinguish between the ruggedness and the robustness of analytical methods. As well as defining the types of test to be conducted, the description of ruggedness testing in the Canadian Guidelines indicate that the means obtained in inter-analyst or inter-laboratory studies should be within 1% and 2% for raw materials and finished formulations, respectively. However, the acceptance criteria described in the Canadian Guidelines do not consider the precision of the methods used and an alternative approach is to set the acceptance criteria for the qualification of additional laboratories using an appropriate statistical approach such as that described by Westlake [43].

In 1977, Horwitz [44–46] showed that for well-behaved analytical methods the between-laboratory RSD is approximately related to the fraction of analyte in the sample ($f_a$) (Fig. 2.24) by the empirical expression:

$$\mathrm{RSD} \cong Q \cdot e^{(1 - 0.5 \log f_a)} \qquad (2.65)$$

where the pre-exponential constant, Q has a value of 2 for the analysis of trace substances in food [44, 46]. Using a value of Q=2 predicts that the between-laboratory RSDs range between 2 and 4% for tablets in which $f_a$ ranges from 0.01 for a tablet weighing 100 mg and containing 1 mg of active substance and 1 for a directly compressed tablet. Horwitz has also shown that eq. 2.65 is valid down to concentrations as low as $f_a=10^{-9}$ (1 ppb) for the analysis of trace substances, such as aflatoxins in food. Figure 2.24 predicts that the between-laboratory RSDs for analysis of substances present at 1 ppb would be approximately 50% [43, 46].

**Table 2.2**
Summary of Canadian Guidelines on Ruggedness Testing*

| Application | Description | Types of data collected |
|---|---|---|
| ***Level 1***<br>All methods | Verification of the basic insensitivity of the method to minor changes in environmental and operational conditions. Demonstration of the reproducibility§ by a second analyst | Stability of test solutions<br>Effects of extraction time on recovery<br>Chromatographic methods:<br>- columns from same manufacturer<br>- for LC: small mobile phase changes (< 5% change in modifier concentration)<br>- for GC: small changes in flow rate and temperature |
| ***Level 2***<br>Methods to be applied at multiple locations or multiple laboratories in a single location | As Level 1 plus verification of the effects of more severe changes in conditions that might arise from transfer of method to second laboratory | Documentation of the effects of changing equipment, e.g.<br>- chromatography column manufacturer<br>- detector<br>- injector<br>- extraction equipment<br>- data system |
| ***Level 3***<br>Methods to be applied at multiple locations or multiple laboratories in a single location using a variety of equipment | Full collaborative study | As for Level 2 but involving multiple laboratories |

*Adapted from ref [42]
§Means to be within 1% for raw materials and 2% for finished pharmaceuticals

In practice the inter-laboratory RSDs for pharmaceutical dosage form analysis are generally in the range 1.0 - 2.5% ($f_a$ = 0.01 - 1.00). Between-day RSD values of between 5 and 15% are typical for the analysis of drugs in biological fluids at concentrations between 1 µg/mL and 10 pg/mL (see Chapter 10). Thus it would appear that a value of Q=1.5 may be more appropriate to describe the relationship between the inter-laboratory RSDs for the analysis of drugs in biological samples and dosage forms. Considering the wide variety of analytes, matrices and analytical methods, it is surprising that such a simple relationship may be used to describe precision and concentration. However, Hall and Selinger [48] have provided the statistical justification for this relationship (eq. 2.65), which they described as the "Horwitz Trumpet" because of its shape.

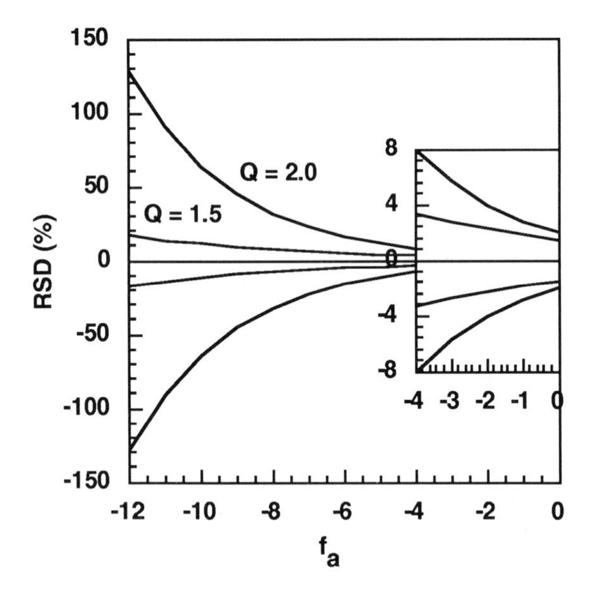


**Figure 2.24**
"Horwitz Trumpet" relating inter-laboratory RSD values to analyte concentration ($f_a$) (eq. 2.65)

### 2.2.8 Analyte stability in the sample matrix

The validation protocol should always include an investigation of the chemical and physical stabilities of the analyte or analytes of interest in the sample matrix as well as the stability of all key components and reagents (including chromatography columns). This is particularly critical if the method is to be automated because a significant period of time may elapse between the time of preparation of test samples and standards and the time when they are actually measured.

Where possible the stability of analytes and key components should be determined early in the development process and the method modified accordingly. Studies to determine the stability of the analyte(s) of interest should be conducted to replicate as closely as possible the actual in-use conditions and the storage period should exceed the expected period of use. This is particularly important for biological samples, which may be stored frozen for several months prior to analysis. For samples stored frozen the stability study should include at least two freeze-thaw cycles [38].

The criteria for the definition of stability should be clearly defined. Suitable criteria for the stability of active substances include: no statistically significant decrease in concentration and the absence of additional impurities over the maximum storage period of the samples and the standards. The criteria for the stability of related compounds expected to be present in trace amount (<1%) can be less restrictive (e.g. the time for 10% degradation). However, metabolites and biotransformation products expected to be present at pharmacologically relevant concentrations should be treated in the same manner as the parent compound. Storage conditions and shelf lives should be defined if any degradation of the analyte(s) of interest is observed. Unstable reagents and key components of the assay should also be labeled with appropriate expiration dates and storage conditions.

## Acknowledgments

The author is particularly grateful to Drs. Earl E. Nordbrock (Anesta Corp), Thomas W. Rosanske (Hoechst Marion Roussel), Martha A. Kral (Hoechst Marion Roussel) and Paul K. Hovsepian

(DuPont Merck) for their helpful advice and suggestions during the preparation of this chapter.

## References

1. J.C. Miller and J.N. Miller, in *Statistics for Analytical Chemistry*, Ellis Horwood, Chicester, p. 7, (1993)

2. J.C. Miller and J.N. Miller, Statisitics for Analytical Chemsits. 3rd ed. New York, Ellis Horwood. (1993)

3. R. Caulcutt and R. Boddy, Statistics for Analytical Chemists. New York, Chapman and Hall. (1983)

4. R. Sokal and F.J. Rohlf, Biometry. The Principles of Statistics in Biological Research. New York, Freeman. (1981)

5. S. Dowdy and S. Wearden, Statistics for Research. New York, Wiley. (1983)

6. *812 Fed. Suppl.*, 458-492, (1993)

7. J.C. Miller and J.N. Miller, in *Statistics for Analytical Chemistry*, Ellis Horwood, New York, p. 18, (1993)

8. M. Hueuermann and G. Blaschke, *J. Pharm. Biomed. Anal.*, **12**, 753-760, (1994)

9. Z. Yu and M.P.C. Ip, *J. Pharm. Biomed. Anal.*, **12**, 787-793, (1994)

10. E. Debesis, J.P. Boehlert, T.E.Givand, and J.C. Sheridan, *Pharm. Tech.*, **Sept.**, 120-137, (1982)

11. M.T. Akermans, F.M. Everaerts, and J.L. Beckers, *J. Chromatogr.*, **549**, 349-355, (1991)

12. D. Demorest and R. Dubrow, *J. Chromatogr.*, **559**, 43-56, (1991)

13. K.D. Altria, *Chromatographia*, **3**, 177-182, (1993)

14. J.N Miller, *Analyst*, **116**, 3-14, (1991)

15. L. Aarons, *J. Pharmacol. Meth.*, **17**, 337-346, (1978)

16. L. Aarons, *Analyst*, **106**, 1249-1254, (1981)

17. L. Aarons, *J. Pharm. Biomed. Anal.*, **2**, 395-402, (1984)

18. J.L Burrows and K.V. Watson, *J. Pharm. Biomed. Anal.*, **12**, 523-531, (1994)

19. M.F. Delaney, *J. Liq. Chromatogr.*, **3**, 164-268, (1985)

20. J.S. Garden, D.G. Mitchell, and W.N. Mills, *Anal. Chem.*, **52**, 2310-2315, (1980)

21. H.T. Karnes and C. March, *J. Pharm. Biomed. Anal.*, **9**, 911-918, (1991)

22. A.M. McLean, D.A. Ruggirello, C. Banfield, M.A. Gonzalez, and M. Bialer, *J. Pharm. Sci.*, **79**, 1005-1008, (1990)

23. P.K. Wilkinson, J.G. Wagner, and A.J. Sedman, *Anal. Chem.*, **47**, 1506-1510, (1975)

24. J. Wolters and G. Kateman, *J. Chemometrics*, **3**, 329-342, (1989)

25. L.J. Phillips, J. Alexander, and H.M. Hill, *Method. Surveys Biochem. Anal.*, **20**, (1990)

26. J. Smisterova, K. Ensing, and R.A. de Zeeuw, *J. Pharm. Biomed. Anal.*, **12**, 723-745, (1994)

27. K.A. Connors, in *A Textbook of Pharmaceutical Analysis*, John Wiley, New York, p. 423, (1975)

28. N.A. Guzman, J. Moschera, K. Iqbal, and E.W. Malick, *J. Chromatogr.*, **608**, 197-204, (1992)

29. *British Pharmacopoeia*, Her Majesty's Stationery Office, London, A82-A86, (1988)

30. *European Pharmacopoeia*, Maisonneuve, Sainte-Ruffine, V.6.20.4, (1987)

31. *European Pharmacopoeia*, Maisonneuve, Sainte-Ruffine, V.6.20.3-1, (1992)

32. *United States Pharmacopeia*, United States Pharmacopeial Convention, Rockville, 1774-1779, (1995)

33. J.P. Foley and J.G. Dorsey, *Anal. Chem.*, **55**, 730-737 (1983)

34. S. Chishima, *J. Pharm. Biomed. Anal.*, **12**, 795-804, (1994)

35. *United States Pharmacopeia*, United States Pharmacopeial Convention, Rockville, 1982-4, (1995)

36. J.C. Wahlich and G.P. Carr, *J. Pharm. Biomed. Anal.*, **8**, 619-623, (1990)

37. *Spectrochim. Acta B*, **33**, 242, (1978)

38. V.P. Shah, K.K. Midha, S. Dighe, I.J. McGilveray, F.P. Skelly, A. Yacobi, T. Layloff, C.T. Viswanathan, C.E. Cooke, R.D. McDowall, K.A. Pittman, and S. Spector, *Pharm. Res.*, **9**, 588-592, (1992)

39. G.P. Carr and J.C. Wahlich, *J. Pharm. Biomed. Anal.*, **8**, 613-618, (1990)

40. E.B. Sheinin, *Pharm. Tech.*, 82-96, (1986)

41. *Committee for Proprietary Medicinal Products: Validation of Analytical Procedures*, Commission of the European Communities, Brussels, (1994)

42. *Drugs Directorate Guidelines*, Health Protection Branch, Health and Welfare, Canada, Ottawa, 1-35, (1992)

43. W.J. Westlake, *Biometrics*, **32**, 741-744 (1976)

44. W. Horwitz, *J. Assoc. Off. Anal. Chem.*, **60**, 1355-1363, (1977)

45. T. Layloff, *Referee*, **Oct.**, 1-8, (1991)

46. T. Layloff and P. Motise, *Pharm. Tech.*, **Sept.**, 122-132, (1992)

47. W. Horwitz, L.R. Kamps, and K.W. Boyer, *J. Assoc. Off. Anal. Chem.*, **63**, 1344-1354, (1980)

48. P. Hall and B. Selinger, *Anal. Chem.*, **61**, 1465-1466, (1989)

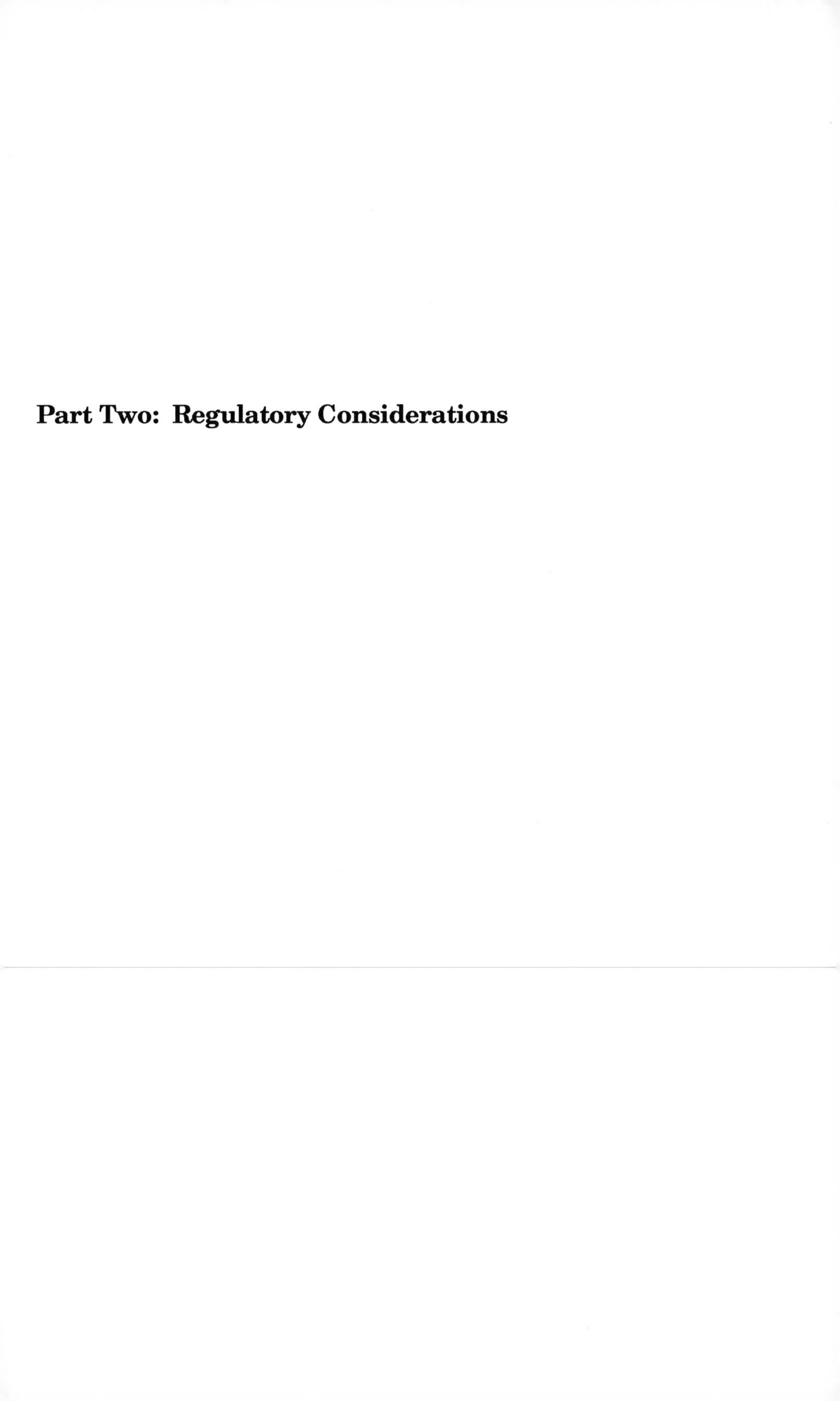

# Part Two: Regulatory Considerations

# Chapter 3

# Overview of World-wide Regulations

*Ian E. Davidson*

## 3.1 Introduction

Analytical validation is a very important feature of any package of information submitted to international regulatory agencies in support of new product marketing or clinical trials applications. The extent of detailed guidelines for requirements provided by different agencies varies widely.

This chapter will review the guidelines provided by the three main geographical regions, the USA, Europe and Japan, for the validation of methods used in pharmaceutical analysis and bioanalysis. Microbiological methods are excluded since definitive guidelines do not exist in most regions. Regulations from some of the smaller regulatory agencies such as Canada and Australia will also be mentioned, since these agencies provide some of the most clear directions on method validation. Some regions have had their own independent guidelines for many years. The recent International Conference on Harmonization (ICH) initiative, which is described in this chapter, is in the process of attempting to harmonize several regulatory guidelines in the area of pharmaceutical drug substance and finished product testing, including one for analytical method validation.

The first part of this chapter will review the individual regional guidelines for drugs and pharmaceutical products prior to the ICH process, followed by a description of the harmonized proposals of ICH. Current Good Manufacturing Practices (cGMPs) will be mentioned, and reference made to the recent Barr case in the United States, which is discussed in more detail in Chapters 4 and 5.

Subsequent sections will discuss requirements for the validation of bioanalytical methods. Generally, guidelines are given for product marketing applications, rather than clinical trial submissions. However, the same general principles can be used in both situations.

## 3.2 Pharmaceutical-analysis methods used to control the quality of drug substances and medicinal products

This section deals with the analysis of drug substances and products from the standpoint of determining the quality of the materials tested. The regulatory agencies have concentrated on assay and impurity methods for drug substances and products, identity tests, and dissolution methods. Chromatographic procedures are among those most frequently used in pharmaceutical analysis and have been the subject of the most extensive work in method validation.

### 3.2.1 The European Community guidelines

The regulatory process in Europe has historically been rather complex. Until recent years, each country had its own registration requirements. With the advent of the European-wide Committee for Proprietory Medicinal Products (CPMP) process for product approvals, the European Community (EC) has issued guidelines intended to cover all countries in the community. The present discussion focuses on the current European requirements, rather than attempting to describe previous differences between individual countries.

In July 1989 [1] and then in 1992 [2], the EC issued an analytical validation guideline in their publication "The Rules Governing Medicinal Products in the European Community". The European guidelines on analytical method validation have also been summarized by Cartwright and Matthews in a recent book [3].

Analytical validation is summarized as including tests to confirm specificity, precision (both repeatability and reproducibility), accuracy, linearity, range, sensitivity, limit of detection, limit of quantitation, and robustness. Which of these needs to be established should be decided on a case by case basis. The definitions of each of these parameters will be given in the section below on the ICH guidelines (see also Chapter 2).

It is important to note that the European guidelines, in common with the other major international regions, have never given explicit details for number of replicate determinations, analysts, instruments, days of testing, etc. The European guidelines [1, 2] indicate that they are applicable to the following sections of the chemical, pharmaceutical and biological documentation (the numbers refer to the relevant CPMP submission sections):

11-A Development pharmaceutics

11-B In-process control during manufacturing process

11-C Control of the starting materials (active ingredients-other components if necessary)

11-D Control tests on intermediate products

11-E Control tests on the finished products

11-F Stability

The guidelines state that revalidation of the procedure may be necessary in certain circumstances such as transfer from analytical development to quality control, or when significant changes in the manufacturing process of the starting material or in the composition of the finished product have occurred. The degree of revalidation depends on the nature of the changes. Aspects to be validated depend on the procedure and do not apply in all circumstances, but could include:

a) Identification: Specificity

b) Impurity content tests: Specificity, limit of detection (LOD) or limit of quantification (LOQ)

c) Assay (Content or Potency): Specificity, repeatability, precision, reproducibility, accuracy, linearity, range and sensitivity

General recommendations are given which require that the procedures include method principles, and are described in such a way that they can be repeated by regulatory authority or state laboratories. The description should include test conditions, precautions, reagents, reference substances and preparations.

In a chromatographic system, a system suitability test should be provided. Detailed formulas for result calculations should be given, together with precise descriptions of equipment not commercially available. In this case details of a method as similar as possible using standard equipment should be given. Methods found in pharmacopoeias are considered to be validated, provided they are used for the intended application. Similarly, reference substances should be evaluated for their intended purpose. Complete data showing validity should be indicated.

Overall, the validation guidelines are very general, with little practical guidance, and companies are free to use scientific judgement as to the extent of validation required.

### 3.2.2 The United States Food and Drug Administration.

In February 1987, the US Food and Drug Administration (FDA) issued a document entitled "Guideline for Submitting Samples and Analytical Data for Methods Validation" [4]. For New Drug Applications (NDAs) in the United States, the FDA requires that analytical methods are tested in the FDA's own laboratories and that suitable samples are provided for that purpose. These guidelines describe the types of materials to be submitted, contents of the methods validation package, and information supporting the suitability of the methodology for both drug substance and dosage form. Some definitions are given in the guidelines concerning regulatory specifications and methodology.

Regulatory methods validation is defined as the process whereby submitted analytical procedures are first reviewed for adequacy and completeness and are then tested as deemed necessary in FDA laboratories. These requirements have been summarized in a book by Swit [5] and may be summarized as follows:

**a. Types of material to be submitted** Four identical sets of samples need to be submitted (as described in FDA regulations 21 CFR 314.50(e) [6]) after the FDA chemists have reviewed the validation information provided. Usually the samples will be sent to two FDA testing laboratories. The FDA will normally request standards and unusual reagents or equipment.

**b. The contents of the method validaton package to be submitted to the FDA** will usually contain information copied from relevant sections of the NDA. Detailed information should include the following:

1. Tabular listing of all samples to be submitted
2. Listing of all proposed regulatory specifications
3. Information supporting the integrity of the reference standard
4. A detailed description of each method of analysis
5. Information supporting the suitability of the methodology.

This is decribed below in more detail for both drug substance and drug product.

**c. The validation guidelines** are very general, and are divided into requirements for drug substance and dosage forms.

**d. Information Supporting the suitability of the methodology for a new drug substance** The FDA guidelines state that generally, the following information should be provided:

1. A summary flow chart of the synthesis of the drug, with a list of known impurities and side reaction products
2. Information on control of polymorphs/isomers
3. Data demonstrating suitable accuracy, precision and linearity over a range of about 80% to 120% of theory. Data demonstrating specificity of the methods and determination of limits for degradation products or impurities should be included. These degradation products or impurities should be adequately identified and characterized.

4. Legible reproductions of representative chromatograms and instrumental records

**e. Information supporting the suitability of the methodology for the dosage form** should include:

1. Data demonstrating suitable accuracy, precision, and linearity over a range corresponding to 80% to 120% of the target value (Data demonstrating specificity of the methods and determination limits for degradation products or impurities should be adequately identified and characterized.)

2. Data demonstrating recovery from the sample matrix where the nature of the product so indicates

3. Data demonstrating that neither fresh nor degraded placebo interferes with the proposed method

4. Legible reproductions of representative chromatograms and instrumental recordings

5. Data characterizing day-to-day, laboratory-to-laboratory, analyst-to-analyst, and column-to-column (for chromatographic methods) variability (These data may be included to provide a further indication of reproducibility and, in a limited sense, ruggedness.)

6. A degradation schematic for the active ingredient in a dosage form, where possible (e.g., products of acid/base hydrolysis, temperature degradation, photolysis, and oxidation)

Generally, a stability-indicating method will be employed as the regulatory method. If the assay is not stability-indicating, then limit tests for degradation products must be submitted.

The guidelines list a number of common problems that can delay a successful validation by the regulatory agency. These include:

1. Failure to include a suitable sample of an impurity, degradation product or internal standard necessary to assess the adequacy of the method

2. Failure to list complete specifications, or selecting unsuitable specifications, e.g.,unsubstantiated or broad ranges, or specifications that do not account for assay limitations

3. Failure to provide sufficient detail, or unacceptable choice of procedures, reagents or equipment

4. Failure to submit well-characterized reference standards.

The guidelines also list characteristics useful in defining particular chromatographic columns such as packing material, column type and system suitability tests.

### 3.2.3 The United States Pharmacopeia and National Formulary

The USP contains a section on "Validation of Compendial Methods" [7]. According to Section 501 of the Federal Food, Drug and Cosmetic Act, assays and specifications in monographs of the USP and the National Formulary (NF) constitute legal standards. The current Good Manufacturing Practice regulations (21 CFR 211.194(a)) require that test methods, which are used for assessing compliance of pharmaceutical products with established specifications, must meet proper standards of accuracy and reliability. Also, according to these regulations (21 CFR 211.194(a)(2)), users of analytical methods described in the USP and the NF are not required to validate accuracy and reliability of these methods, but merely verify their suitability under actual conditions of use. Recognizing the legal status of USP and NF standards, it is essential that proposals for adoption of new or revised compendial analytical methods be supported by sufficient laboratory data to document their validity. New methods submitted to the USP must contain sufficient data to enable members of the USP Committee of Revision to evaluate adequately the procedures.

The USP describes the typical analytical parameters used in assay validation, i.e., accuracy, precision, specificity, limit of detection, limit of quantitation, linearity and range. Different validation schemes are proposed for different test methods. The common categories of assays covered in the USP are:

| | |
|---|---|
| Category 1 | Methods for quantifying major components of bulk drug substances or active ingredients in finished pharmaceutical products |
| Category 2 | Methods for impurities in bulk drug substances or degradation products in finished products. These include quantitative assays and limit tests |
| Category 3 | Methods for determination of performance characteristics (e.g.dissolution, drug release) |

A tabulation is provided in the USP of the data elements required for the different categories, similar to those described in the ICH guidelines discussed below.

### 3.2.4 The Japanese Ministry of Health and Welfare (MHW)

The Japanese regulations, Drug Approval and Licensing Procedure in Japan 1992 [8], do not give specific guidance on the requirements for analytical method validation. This has been entrusted to the scientific judgement of each individual pharmaceutical company. Validation is mentioned as a matter to be addressed in the setting of specifications and test methods. The guidelines state that "*standards and test methods are carefully investigated when an application is made for manufacturing approval.*" Japanese representatives to the ICH Quality working groups have been very active and the MHW has now issued for review some draft proposed guidelines based on the ICH proposals.

### 3.2.5 The International Conference on Harmonization (ICH). Guidelines on Analytical Method Validation

The ICH process is an important regulatory initiative and it is probably worthwhile giving a brief background before describing its relevance to the regulatory aspects of analytical method validation. The ICH was initiated around 1990, based on the recognition of inconsistencies in the requirements for new drug product registration in the major global markets. These inconsistencies were leading to significant duplication, inefficiency and delay in the introduction of new drug products globally. ICH is concerned with harmonization of technical requirements for the registration of products, among the three major geographical markets of the European Community, Japan and the United States of America. The terms of reference of ICH are:

- To provide a forum for a constructive dialogue between regulatory authorities and the pharmaceutical industry on the real and perceived differences in the technical requirements for product registration in the European Community, USA and Japan

- To identify areas where modifications in technical requirements or greater mutual acceptance of research and development procedures could lead to a more economical use of human, animal and material resources, without compromising safety

- To make recommendations on practical ways to achieve greater harmonization in the interpretation and application of technical guidelines and requirements for registration

ICH is a joint regulatory/industry undertaking and the six co-sponsors of the conference are:

Europe: European Commission (EC) and European Federation of Pharmaceutical Industry Associations (EFPIA)

Japan: Ministry of Health and Welfare (MHW) and the Japanese Pharmaceutical Manufacturers Association (JPMA)

USA: Food and Drug Administration (FDA) and Pharmaceutical Research and Manufacturers of America (PhRMA)

Some organizations, including the Canadian Health Protection Branch and the United States Pharmacopeia act as observers to the conference working committees and their input is also influential in the guideline-setting process. In addition, the International Federation of Pharmaceutical Manufacturers Associations (IFPMA) acts as an "umbrella" organization for the pharmaceutical industry, and provides the "ICH Secretariat" to coordinate the preparation of documentation.

Three separate areas of regulatory requirements were identified for harmonization, namely Safety, Efficacy and Quality, and a working party for each was established to define topics for discussion. Within the Quality section, five topics were selected for initial harmonization, *viz.*,

- Stability Guidelines
  -New chemical entities
  -Extension to light stability

- Analytical Validation

- Guidelines on Impurity Testing

- Pharmacopeial Harmonization

- Quality of Biotechnological and Biological Products
  -Manufacturing Variations
  -Genetic stability
  -Stability of Biotechnology Products

The finalization and issuance of ICH guidelines is a five-step process as described in Table 3.1 In October 1994, the Step 4 draft ICH Guidelines on "Validation of Analytical Procedures: Definitions and Terminology" were endorsed by the three participating regulatory agencies, and are now final and implementable[1].

The ICH Guidelines present a discussion of the characteristics that should be considered during the validation of the analytical procedures included as part of registration applications within Europe, Japan and the United States.

---

[1]Editors' note: Just prior to this book going to press, Step 2 of the ICH Concensus Draft entitled "Validation of Analytical Procedures: Methodology" was issued (November 29, 1995) and was too late to be discussed here.

**Table 3.1.**
Steps in the ICH process

| | |
|---|---|
| Step 1 | Preliminary discussions of the topic are held by a group of experts, mandated by the ICH Steering Committee, and a preliminary draft is prepared. The draft is discussed by the joint industry/regulatory ICH Expert Working Group. Usually this process results in numerous drafts and iterations of a guideline before concensus is achieved. |
| Step 2 | The draft is transmitted to the three regional regulatory agencies for formal consultation inaccordance with their internal consultation procedures. |
| Step 3 | The draft is amended, if necessary, in the light of the consultation process. The revised draft is discussed by the joint industry/regulatory ICH Expert Working Group. |
| Step 4 | The final draft is endorsed by the Steering Committee which recommends adoption by the three regulatory bodies. |
| Step 5 | The recommendations are incorporated into domestic regulations or other appropriate administrative measures, according to national/regional internal procedures. |

The discussion is directed to the four most common types of analytical procedures:

- Identification tests
- Quantitative measurements for impurities
- Limit tests for the control of impurities
- Quantitative measurement of the active moiety in samples of drug substance or product or other selected components in the drug product.

Tests such as dissolution may be addressed in other ICH documents, but are excluded from the one on Analytical Validation. A brief description of the types of tests considered in the ICH proposals are given, together with a glossary of terms and definitions for: analytical procedure, specificity, accuracy, precision (repeatability, intermediate precision, reproducibility), detection limit, quantitation limit. Table 3.2 describes the type of analytical procedure and characteristics for the different types of test.

### 3.2.6 The Canadian Health Protection Branch

The Canadian Health Protection Branch (HPB) issued a Drugs Directorate Guideline in 1992 entitled "Acceptable Methods" [9]. The guideline covers basic principles for methods, validation and revalidation, methods content and documentation, reference standards and performance checks. Chromatographic, dissolution and physicochemical methods are covered. Typical parameters that should be covered in validation are similar to those already mentioned, but specific definitions and procedural guidelines are given as follows:

Precision - Two measures are required - precision of the method and of the system. The method precision should be performed a) intra-day, with six assays of one drug substance or six assays of a composite of the drug product on a single day - the relative standard deviation should be less than 1.0% for drug substance and less than 2.0% for dosage forms, and b )inter-day - duplicate or triplicate assays on five consecutive days. For system precision of chromatographic methods, five replicate injections of a standard solution should give a coefficient of variation of less than 1.0%. Precision limits for impurities/related substances are also provided.

Accuracy - The required accuracy is a bias of 2% or less for dosage forms, and 1% or less for drug substances.

Limit of detection - For instrumental methods, a signal to noise ratio of 3:1 is generally accepted.

Limit of quantitation - Defined for instrumental methods as the best estimate of the lowest concentration of an analyte that gives a relative standard deviation of approximately 10% for six replicate determinations. The method should be able to quantitate each individual impurity for which the method will be applied at or below the limit for that impurity.

Selectivity/specificity - Selectivity is determined by comparing test results from the analysis of samples containing impurities, degradation products, or placebo ingredients, with those obtained from the analysis of samples without them.

Sensitivity - This is defined as the slope of the linear regression line (see also Chapter 2).

Linearity and range - Linearity is determined by mathematical treatment of test results obtained by analysis of samples with analyte concentrations across the claimed range of the method. The range of the method is estimated from the linearity and is verified by confirming that the method provides acceptable precision and linearity when applied to samples containing analyte at the extremes of the range as well as within the range.

Ruggedness - Determined by analysis of aliquots from homogeneous lots in different laboratories, by different analysts (see also Chapter 2).

Table 3.3 shows the data required the Canadian Guidelines for different types of analysis.

**Table 3.2**
Tabulation Of Validation Requirements For Various Types Of Test (ICH Guidelines)

| Type of analytical procedure: characteristics | Identification | Testing for impurities | | Assay: dissolution (measurement) only Content/potency |
|---|---|---|---|---|
| | | Quantitation | Limit | |
| Accuracy | - | + | - | + |
| Precision: | | | | |
| Repeatability | - | + | - | + |
| Intermediate precision | - | +[1] | - | +[1] |
| Specificity[2] | + | + | + | + |
| Detection limit | - | -[3] | + | - |
| Quantitation limit | - | + | - | - |
| Linearity | - | + | - | + |
| Range | - | + | - | + |

Note: - signifies that this parameter is not normally evaluated; + signifies that this parameter is normally evaluated; [1] in cases where reproducibility (see Glossary in ICH Guidelines) has been performed, intermediate precision is not needed; [2] lack of specificity of one analytical procedure could be compensated by other supporting analytical procedure(s); [3] may be needed in some cases

**Table 3.3**
Canadian HPB Guidelines for Parameters for Assay Validation

| Analytical Performance Parameters | Group 1 Identity Tests | Group 2 Active Ingredients | | Group 3[a] Impurities/ Degradation Products | | Group 4 Physico-chemical Tests |
|---|---|---|---|---|---|---|
| | | Bulk Drug Substance | Drug Product | Quanti-tative | Limit Tests | |
| Precision: of the system | No | Yes | Yes | Yes | No | Yes |
| Precision: of the method | No | * | Yes | Yes | * | Yes |
| Accuracy | No | * | Yes | Yes | * | Yes |
| Limit of detection | * | No | No | No | Yes | * |
| Limit of quantitation | No | No | No | Yes | No | * |
| Selectivity/ specificity | Yes | * | Yes | Yes | Yes | * |
| Range | No | * | Yes | No | No | Yes |
| Linearity | No | Yes | Yes | Yes | No | Yes |
| Ruggedness | * | Yes | Yes | Yes | Yes | Yes |

[a]"Quantitative" refers to tests which result in a quantitative estimate of the concentration of a specific impurity, while "limit tests" refers to tests which indicate whether an impurity is present above or below a specified value.
* May be required depending on the nature of the test

The Canadian guidelines also describe the requirements for revalidation. Tests should be revalidated to confirm their accuracy when used over a long period of time, for example one year, or if significant changes are made to the equipment or conditions of analysis, or if the drug product or material being analyzed has changed.

### 3.2.7 Current Good Manufacturing Practice Guidelines

Analytical method validation is an essential component of adherence to current Good Manufacturing Practice guidelines (cGMPs). To quote the US FDA's cGMP guidelines [10] - "*The accuracy, sensitivity, specificity and reproducibility of test methods employed by the firm shall be established and documented*" (Section 211.165), and laboratory records shall include complete data derived from all tests necessary to assure compliance with established specifications and standards, including examinations and assays, as follows: "*A statement of each method used in the testing of the sample: The statement shall indicate the location of data that establish that the methods used in the testing of the sample meet proper standards of accuracy and reliability as applied to the product tested.........The suitability of all testing methods shall be verified under actual conditions of use*" (Section 211.194).

Further discussion of cGMPs is beyond the scope of this chapter. More information on the legal interpretation of the FDA guidelines will be found in Chapter 4 which details the recent case involving the United States *versus* Barr Laboratories Inc.

## 3.3 Methods used in the analysis of biological samples

### 3.3.1 Introduction

Bioanalytical methods are defined here as chemical methods used to measure the levels of drugs and metabolites in biological fluids such as plasma, serum, urine and cerebrospinal fluid, etc. (see also Chapter 8). Measurements are usually taken to gain information on drug pharmacokinetics, product bioequivalence, drug in-vivo interactions and the effect of such parameters as disease state, age and fasting versus feeding, on drug absorption. The validation requirements are somewhat different from those in pharmaceutical analysis since the levels of drug being measured in biological fluids are generally much lower than in pharmaceutical products, normally in the pg/ml to µg/ml range. Furthermore, bioanalytical materials are often more complex, involving more extensive sample preparation than is necessary for the analysis of a drug substance or a finished dosage form. Consequently, the acceptance criteria for accuracy and precision levels are somewhat more generous, although the methods must still be sufficiently accurate and precise to meet the requirements of the data to be generated from the analysis.

Bioanalytical studies are applied in both the preclinical animal study and human clinical study phases of drug development. Validated analytical methods are required in each case. The extent of guidance provided by the regulatory agencies varies considerably, and is not as well-defined as in pharmaceutical analysis.

### 3.3.2 The European Community

An EC guideline [11] provides basic guidance on the presentation of data on the validation of test procedures carried out for toxicological and pharmacological studies as well as for clinical trials provided for by Directive 75/318/EEC.

The objective of analytical validation on samples of biological origin such as plasma and urine is to demonstrate the reliability of results for drug and metabolites obtained from pharmacokinetic, metabolic and bioavailability studies. The criteria are those contained in the Good Laboratory Practice (GLP) guidelines and consist of: specificity, precision (repeatability/reproducibility), accuracy, linearity, range, sensitivity, limit of detection and limit of quantification.

No specific details are given for how validation should be performed, but several recommendations are provided. Since bioanalysis is often carried out in more than one laboratory, it is very important to be able to compare results between laboratories. Two cases should be considered:

1. When the same test procedure is always used, quality control between the laboratories is necessary; and

2. when using different test procedures, either a reference test method (control, standard) or an investigation of recovery in the individual methods using the same reference material should be available.

Stability of the drug in the sample matrix should be determined for the length of time the matrix will be stored before testing, and studies on adsorption by the sample container and stopper should be undertaken. Test procedures must be described precisely, including the mode of sampling, conditions of storage before analysis, and an exact description of the test conditions, including precautions, methods of extraction, reagents, reference substances and preparations, should be provided.

A system suitability test should be included to verify the separating power of chromatographic systems. Detailed formulas for the calculation of results should be provided, as well as statistical evaluations where appropriate. Reference substance identity, purity and content must be fully established.

### 3.3.3 The United States Food and Drug Administration

The FDA provides a guideline for the format and content of the human pharmacokinetics and bioavailability section of an NDA application [12]. This does not give specific guidance on the validation of analytical methods, except to say that a summary of the analytical methods employed should be provided together with documentation on the sensitivity, linearity, specificity and reproducibility of the analytical method including sample chromatograms and recovery studies.

In 1990, a conference was held on the subject of "Analytical Methods Validation: Bioavailablity, Bioequivalence and Pharmacokinetic Studies", sponsored by the American Association of Pharmaceutical Scientists, the US Food and Drug Administration, Federation International Pharmaceutique, the Health Protection Branch (Canada) and the Association of Official Analytical Chemists. (see also Chapter 8.) The report from the conference was published

[13] and was intended to provide guiding principles for validation of methods used in bioavailability, bioequivalence and pharmacokinetic studies. The report was intended to provide a framework for the future development of US guidelines on the subject. Some principles and requirements for establishing a valid method are described and include:

1. A specific detailed description of the method should be provided.

2. Each step should be investigated to determine the extent to which matrix and environmental variables could affect the determination of the analyte.

3. A method validation report should be provided.

4. The same biological matrix should be used for validation as in the intended real samples. The stability of the drug in the matrix should be determined.

5. The concentration range must be defined in the method, and a standard curve derived.

6. An adequate number of standards must be used to define adequately the relationship between concentration and response.

7. The accuracy and precision with which known concentrations of analyte in the biological matrix can be determined must be demonstrated. Within-run and between-run accuracy and precision should be calculated with commonly accepted statistical procedures. Specific criteria must be set for accuracy and precision over the range of the standard curve.

8. The limit of quantitation (LOQ) should be determined by using at least five independent samples of standards and by determining the coefficient of variation and/or appropriate confidence intervals.

In addition to principles, some specific recommendations should be considered when validating a method:

1. The stability of the analyte in the biological matrix at the intended storage temperature(s) should be established.

2. The specificity of the assay method should be established with six independent sources of the same matrix.

3. The accuracy and precision should be determined with a minimum of five determinations per concentration. The mean value should be within +/- 15% of the actual value, except at LOQ, where it should not deviate by more than +/- 20%. The relative standard deviation around the mean value should not exceed 15%, except for the LOQ where it should not exceed 20%.

4. The standard curve should consist of five to eight standard points, excluding the blank, with single or replicate samples. The standard curve should cover the entire range of expected concentrations (see also Chapter 2).

5. The simplest relationship for response versus concentration should be determined, and the fit should be statistically tested.

This published proceedings of the conference gave acknowledgement to a very wide list of journals, individuals and organizations including the FDA and the Health Protection Branch of Canada. As stated earlier, the FDA does not give detailed instructions for method validation. However, field inspectors conduct audits of bioequivalence and bioavailablity data, through district compliance branches. Companies therefore need to be conducting adequate validation studies which confirm the validity of data submitted in support of conclusions about bioavailability and bioequivalence.

The FDA inspectors are provided with checklists from which to compile audit reports. For bioequivalence testing, these include descriptions of the testing organization's facilities, personnel, specimen handling procedures, equipment, prestudy analysis, in vivo methods validation and sample analyses, data handling and storage.

Some of the major non-compliances which inspectors have found include:

1. Improper storage of samples

2. Unskilled personnel conducting analyses

3. Poor laboratory techniques

4. No records to document findings

5. Failure to employ standard quality control techniques such as standard curves

6. Absence of equipment the laboratory claims to be using

7. Significant non-agreement between data reported to FDA and those audited on site

8. Lack of validation of assay methodology

Less serious issues include:

1. Inadequate records of study sponsors

2. Inadequate standard operating procedures

3. Lack of written procedures for sample handling

### 3.3.4 The Japanese Ministry of Health and Welfare(MHW)

As for pharmaceutical products, the MHW does not issue specific guidelines for method validation in biological matrices. In the Japanese Guidelines for Nonclinical Pharmacokinetic Studies [14], the objectives of the studies are intended to examine the absorption, distribution, metabolism and excretion of drugs, and issues of test methods and parameters to be determined are discussed. The only reference to validation appears in the test method section, where it states "*Assay methodology: The assay method and its sensitivity, precision, specificity, etc. should be clearly defined*".

### 3.3.5 The Australian Therapeutic Goods Administration

The Therapeutic Goods Administration (TGA) issued quite detailed "Guidelines for Bioavailability and Bioequivalence Studies" in 1989 [15]. The guidelines contain details of assay validation requirements. Preference should be given to "state of the art" HPLC or GC methods, and assay validation must be conducted in the laboratory which generated the study data. Detailed data on the specificity, accuracy, reproducibility and sensitivity of the analytical procedure must be supplied as part of the study report.

Specificity - Evidence must be provided that the assay does not have interference from endogenous compounds, metabolites or other drugs likely to be present in the study samples.

Stability - Data should be accumulated to show the stability of the drug or metabolites in the relevant biological environment from time of sampling to assay, under the conditions and duration of storage that apply.

Minimum quantifiable concentration (MQC) - The TGA defines MQC as the lowest concentration which has a relative standard deviation of 20%, as best this can be measured. It is not necessary to define an MQC if the lowest concentration encountered during sampling had a relative standard deviation of less than 20% during assay validation.

Shape of the calibration curve - Linearity is preferred, and the shape should be defined in mathematical terms more than once (preferably three times), over a concentration range from the MQC to a value greater than the $C_{max}$ expected in the study. Calibration standards in the 1,2,5,10,20,50---- pattern of concentrations are preferable (see also Chapter 2). The coefficient of determination ($r^2$) should normally exceed 0.99.

Assay precision and accuracy should be documented at three concentrations (low, medium and high) where low is in the vicinity of the lowest concentration to be measured, high is a value in the vicinity of $C_{max}$ and the medium is a suitable intermediate value. Intra-assay precision (within days) in terms of coefficient of variation should be no more than 10% although no more than 20% may be more realistic at values of the order of the MQC. Inter-assay precision (between days) may be higher than 10% but not more than 20%.

Accuracy can be assessed in conjunction with precision and in general an accuracy of +/- 10% should be obtained.

Recovery - Documentation of extraction recovery at high and low concentrations is essential since methods with low recovery are in general more prone to inconsistency. If recovery is low, alternative methods should be investigated.

From the standpoint of assaying study samples, calibration standards and a sample blank (e.g. plasma) should be analyzed with each batch of study samples on a daily basis. Calibration standards should be blank samples (e.g. plasma) spiked with known concentrations of drug and prepared freshly each day from pure reference substance.

Standards and seeded controls should be spaced throughout the batch. Failure to obtain reproducibility and linearity for the daily standards necessitates re-assay of the batch. Assays should be repeated if more than one control lies outside two standard deviations (inter-assay) of the mean for that concentration.

Seeded controls - These are a valuable component of in-study quality assurance. Controls at three or more concentrations are prepared in plasma in bulk at the time of pre-study assay validation, or at the time of sample collection. A control for each concentration is assayed on each occasion that the study samples are analysed, and the concentration determined by reference to that day's calibration standards. If the concentration values determined from the seeded controls are not within ±10% of the expected concentrations, the batch should be considered for re-analysis. If not within ±20% of the expected concentrations, the batch must be re-analyzed unless there is a very good reason not to do so. Seeded controls therefore provide a constant reference point between batches of assays as well as determining whether the drug is stable under the storage conditions used.

Range of reported values - Concentration values less than the lowest calibration standard should be reported as "less than the lowest calibration standard" and should not normally be used in data analysis. Concentration values greater than the highest concentration used in the linearity studies should be highlighted and accompanied by validation data. Concentration values marginally above daily calibration standards may be reported if pre-study linearity data extended beyond such values. Concentration values less than the MQC must not be used in data analysis.

The TGA gudelines also cover validation requirements for radioimmunoassays.

### 3.4 Summary

The extent of guidance on method validation for both pharmaceutical analysis and bioanalysis currently varies widely from region to region. With the increasing communication between regional authorities and the influence of ICH, this is expected to change and it is likely that consistent guidelines will be developed across each region. If the guidelines are scientifically sound, reasonable and agreed to by both industry and the regulators, this should be of value to both partners in the drug development and approval process.

### References

1. *The Rules Governing Medicinal Products in the European Community*, Volume 3 Addendum, July 1990, Catalogue number CB-59-90-936-EN-C

2. Commission of European Communities-CPMP Working Party on Quality of Medicinal Products-"*Specifications and Control; Tests on the Finished Product*" (1992)

3. A.C. Cartwright and B.R. Matthews (eds.), *Pharmaceutical Product Licensing Requirements for Europe*, Ellis Horwood, London, p. 61 (1991)

4. *Guidelines for Submitting Samples and Validation Data for Methods Validation,* US Food and Drug Administration, Center for Drugs and Biologics, Department of Health and Human Sevices, (1987)

5. *Getting Your Drug Approved: FDA's Own Guidelines*, Second Edition, Published by Washington Business Information Inc. (1990)

6. *Federal Register*, **21**, *CFR*, 314.50(e) (199).

7. *United States Pharmacopeia*, United States Pharmacopeial Convention, Rockville, p 198 (1995)

8. *Drug Approval and Licensing Procedures in Japan*, Japanese Ministry of Health and Welfare (1992)

9. *"Acceptable Methods" - Drug Directorate Guidelines*, National Health and Welfare, Health Protection Branch, Health and Welfare Canada (1992)

10. *Current Good Manufacturing Practice for Finished Pharmaceuticals (cGMP's)*, US Food and Drug Administration, Fifth Revision, (1985)

11. Commission of the European Communities, *"Guidelines on the Quality, Safety and Efficacy of Medicinal Products for Human Use"*, (1990)

12. *Guideline for Submitting Samples and Analytical Data for Methods Validation*, U.S Food and Drug Administration, Center for Drugs and Biologics, Department of Health and Human Services (1987)

13. V.P. Shah, K.K. Midha, S. Dighe, I.J. McGilveray, J.P. Skelly, A. Yacobi, T. Layloff, C.T. Viswanathan, C.E. Cook, R.D. McDowall, K.A. Pittman, S. Spector, *J. Pharm. Sci.*, **81**, 309-312 (1992)

14. *Guidelines for Nonclinical Pharmacokinetic Studies*, Japanese Ministry of Health and Welfare, (1992)

15. *Guidelines for Bioavailability and Bioequivalence Studies*, Drug Evaluation Branch, Therapeutic Goods Administration, Australia, (1989)

# Chapter 4

# Issues Related to the United States v. Barr Laboratories, Inc.

*Cathy L. Burgess, esquire*

## 4.1 Introduction

On February 5, 1993, the United States District Court for the District of New Jersey issued its opinion in United States v. Barr Laboratories, Inc., [1] the FDA's suit against Barr for alleged violations of current Good Manufacturing Practice (cGMP) regulations. As the opinion demonstrates, the fundamental issue in the case was not so much whether Barr had violated cGMP but what conduct cGMP required. Because the cGMP regulations are vague, the court was required to base its decision on the evidence presented and not necessarily on an understanding of science. The result was a lengthy opinion that has generated considerable debate in the pharmaceutical community.

## 4.2 The Trial

The Barr trial consumed fifteen days between August 17 and October 12, 1992. The government produced two witnesses – Compliance Officer David Mulligan, who had inspected Barr on several occasions, and Dr. Robert Gerraughty, an outside expert. Barr produced four expert witnesses – Dr. Sanford Bolton (statistics), Dr. Christopher T. Rhodes (cGMP), Dr. Norman Atwater (analytical chemistry), and Dr. Murray Cooper (microbiology). Barr also produced four Barr employees as witnesses: the Director of Quality Control, the Director of Quality Assurance, the Manager of Special (cGMP) Projects, and the Vice President of Human Resources. In response to the government's allegations, Barr argued that the requirements that FDA sought to impose were neither required nor contemplated by the Food, Drug, and Cosmetic Act or its implementing regulations [2]. Barr also argued that its laboratory and manufacturing practices were consistent with industry standards [3]. The evidence presented consisted of over 2300 pages of testimony, almost 400 documents, and numerous written declarations [4]. The Court was left with the task of sifting through all the evidence and determining the appropriate standards to apply.

## 4.3 The Dispute: What Constitutes cGMP

### 4.3.1 Legal Requirements

Section 501 of the Federal Food, Drug, and Cosmetic Act provides that a drug is adulterated "*if the methods used in, or the facilities or controls used for, its manufacture, processing, packing or holding do not conform to or are not operated or administered in conformity with current good manufacturing practice to assure that such drug meets the requirements of this chapter as to safety and has the identity and strength, and meets the quality and purity characteristics, which it purports or is represented to possess*" [5].

The regulations governing cGMP for drug manufacturers are codified in Title 21, Code of Federal Regulations, Parts 210 and 211.

At trial, FDA and Barr offered quite different interpretations of what the regulations require. FDA argued that "*cGMPs are those manufacturing practices and controls which are currently observed by the better firms in the industry*" [6]. FDA also argued that to the extent that the regulations are silent on a point, manufacturers are required to use FDA guidelines, as well as scientific literature, to determine what constitutes cGMP [6]. By contrast, Barr argued that the cGMP regulations are "*general in nature and subject to considerable interpretation . . . . There are many ways to satisfy the objectives of cGMP, and a firm cannot fairly be said to be in violation of cGMP so long as that firm's approach is rational and not violative of any general principles of cGMP*" [3].

The Court agreed with Barr that the cGMP regulations set the parameters for compliance, allowing for "*conflicting, but plausible*" interpretations of what constitutes cGMP [7]. The Court also agreed with FDA, however, that to the extent the regulations are vague, companies must look to scientific literature or "*employ scientific judgment where appropriate*" [7].

### 4.3.2 Failure Investigations

#### 4.3.2.1 What is a "Failure"?

Perhaps the most contentious issue in the case was the definition of the word "*failure*". Had the Court accepted the government's definition, the U.S. pharmaceutical industry as a whole could not have survived. FDA's initial position was that any out-of-specification result, even in an early stage of a multistage test procedure, is a "*failure*" requiring rejection of the entire production batch [8]. As the trial wore on, FDA retreated from this initial position. For example, FDA agreed that an out-of-specification result at the first stage of a content uniformity test (prescribed by the U.S. Pharmacopeia (USP) as a two-stage test) or at the first or second stage of a dissolution test (a three-stage USP test) is not a "*failure*" [9]. The government also acknowledged that an out-of-specification result attributable to laboratory error is not a "*failure*" [10].

Barr argued that an initial out-of-specification test result is not necessarily a failure. If an out-of-specification result is due to an identified error, it can be invalidated and the test repeated. Even when an out-of-specification result cannot be attributed to an identified laboratory error, Barr argued, it can be overcome through statistical analysis or by retesting [11].

The Court stated that the term "*out-of-specification result*" referred to "*an individual test value that does not meet predetermined specifications*" [12]. Because an out-of-specification result could be attributed to laboratory error or could be discredited through the use of an outlier test or appropriate retesting, the Court rejected the government's definition of "*failure*" [13].

#### 4.3.2.2 Outliers

Barr argued that the use of outlier tests to determine whether a test result is statistically consistent with other test results is "*scientifically reasonable*" [14]. Barr noted that the cGMP regulations require that "*statistical quality control criteria . . . include appropriate acceptance levels and/or appropriate rejection levels*" [15]. What is "*appropriate,*" Barr argued, is a matter of scientific judgment

[14]. FDA challenged Barr's use of statistical outlier tests, calling it "*inappropriate and statistically invalid*" [16].

Although the Court agreed that outlier tests could be used to invalidate out-of-specification results, the Court restricted the use of outlier testing to microbiological assays [17]. The Court's reasoning was that the USP expressly allows the use of outlier tests for microbiological assays but does not discuss their use for chemical tests [17]. The USP's silence on the latter point, said the court, constitutes a prohibition [17]. Thus, on this issue, FDA prevailed in eliminating outlier tests in the chemical area, and Barr prevailed in preserving outlier tests for microbiological assays.

#### 4.3.2.3 Retesting

Historically, Barr's retest procedure was the following: If an analyst generated an out-of-specification result, Barr's laboratory would order retesting on the same sample by two different chemists [hereinafter Barr's "*two out of three*" practice]. Prior to 1992, Barr averaged the results obtained on retest and used that average as the result of record. In 1992, Barr began including the initial out-of-specification result in the average for a result of record [18]. FDA argued that whether the initial out-of-specification result was discarded or included in the final result was irrelevant: both practices were "*unscientific and unsound*" [19].

FDA faulted Barr's retest procedure on two grounds. First, FDA argued that Barr often conducted retests on different samples (allegedly non-homogeneous parts of the batch), and thus, retesting might obscure variability within a batch [20]. FDA's argument was based on its belief that a "*sample*" is a single twenty tablet grind. If an analyst selected a new group of twenty tablets from the same bottle, FDA argued, that is a "*resample*" [21].

Barr's position was that the entire bottle brought to the laboratory by Quality Assurance, in which tablets or capsules were collected throughout the manufacturing run, constitutes the "*sample*," [22] and that any group of twenty tablets from that bottle is part of the same "*sample*" [22]. Barr argued that "*resampling*" means going to another bottle, selected from retains, that had not been filled throughout the production run [22]. The Court adopted Barr's view:

> *A retest is defined as additional testing on the same sample, and thus it necessarily follows an initial test. An analyst performing a retest takes the second aliquot from either: (1) the sample that was the source of the first aliquot . . . or (2) the larger sample previously collected for laboratory purposes . . . . These procedures are equivalent* [17].

FDA's second criticism of Barr's retest procedure was that Barr substituted retesting for an appropriate failure investigation [23]. FDA took the position that most failures are process-related, and therefore retesting only serves to mask process-related problems [24]. In FDA's view, retesting would be appropriate only where a failure investigation had identified definitively a specific laboratory (and hence non-process-related) error [25].

Barr argued that retesting is an important means of investigating an out-of-specification result [26]. "*Additional testing represents an important means of ascertaining whether an out-of-specification test result is due to analytical or other error unrelated to the quality of the product . . . . In testing for potency within a sample, increased testing heightens the likelihood that the true value will be ascertained*" [26]. With respect to its two-out-of-three approach, Barr produced evidence that other pharmaceutical companies, both "*brand-name*" and generic, use the same method for retesting [27].

The Court found that "*[r]etesting is necessary if a failure investigation indicates that analyst error caused an out-of-specification result*" [17]. Adopting Barr's view, the Court held that as part of an investigation "[a] *retest is similarly acceptable when review of the analyst's work is inconclusive*" [17]. In such cases, retesting "*substitutes for or supplements the original test results*" [17]. The Court determined that the actual number of retests necessary to confirm or reject an out-of-specification result is a matter of scientific judgment [27] However, the Court stated that firms must do "*enough testing*" to reach a conclusion [28]. "*Such a conclusion cannot be based on 3 of 4 or 5 of 6 passing results, but possibly 7 of 8*"[28][1].

---

[1] The Court rejected Barr's two-out-of-three retest procedure even though Barr produced evidence that the procedure was used throughout the industry. "The problems with Barr's retesting procedure are not erased by the protocol's popularity at other firms . . . . [I]ndustry practice while instructive also is subject to the cGMP-compliance test."

Finally, the Court stated that a proper testing procedure must establish in advance a point at which all testing will cease and the results will be evaluated [28].

### 4.3.2.4 Averaging

The Court rejected Barr's view that averaging test results is an acceptable industry practice. Averaging, said the court, masks variability among the data [29]. The Court was particularly concerned about averaging chemical assay test data where some of the initial values are within specification and others are not [29]. In such instances, the average result might be within specification, and the batch released, even though individual values are out of specification [29]. The Court was less concerned about microbiological assays, stating that for microbiological assays, "*the USP prefers an average . . . . . when reaching an ultimate judgment about the product*" [29]. Thus, Barr lost its averaging argument for chemical testing but won the point for microbiological testing.

### 4.3.2.5 Adequacy of Failure Investigations

FDA criticized both the content and the timeliness of Barr's failure investigations [30]. According to FDA, a full scale investigation is required for every out of specification result [31][2]. Barr disputed these claims [32]. In Barr's view, the appropriate scope of a particular investigation depends on the type of problem identified [33]. Barr explained that when an analyst obtained an out-of-specification result, the analyst and the supervisor would review the analyst's calculations and inspect the laboratory instruments used in the procedure as the first step in the laboratory investigation [34]. If the laboratory could identify the cause of the problem at that stage, the laboratory invalidated the out-of-specification result. If the laboratory

---

[2] The government argued that an adequate failure investigation required a written analysis that included: (1) identification of the problem; (2) a discussion of possible causes; (3) proposed corrective actions; (4) identification of other batches that might have been affected; (5) comments from Quality Control and Manufacturing upon completion of any corrective actions; and (6) signatures of employees responsible for the investigation.

could not identify the cause of the problem, the laboratory would order retests by two different analysts [35].

Barr argued that the two-out-of-three retesting constituted the laboratory investigation, and the test results were the only documentation needed [36]. If the test results were inconsistent, the laboratory would conduct an appropriate outlier test and then average all consistent results [37]. If the average of all consistent results was within specification, the lab considered the investigation complete and released the batch [37]. If the average was out-of-specification or if one or both of the retest analysts obtained an out-of-specification result, the laboratory notified Quality Assurance to open a formal investigation [38].

The Court agreed with Barr that the proper scope of an investigation depends upon the issue presented. The Court said that all investigations should be documented fully and conducted within thirty business days. The Court also set certain limits on industry's ability to discredit out-of-specification results by restricting the use of outlier tests and by rejecting Barr's two-out-of-three testing procedure [39].

### 4.3.3 Analytical Method Validation

An "*analytical method*" is a laboratory procedure that measures an attribute of a product, component or raw material [40]. "*Analytical method validation*" is the process of demonstrating that an analytical method is reliable and capable of performing the analysis for which the method has been designed [41]. The validation process consists of experiments to determine certain parameters, including precision, ruggedness, accuracy, selectivity and linearity [42].

Section 211.165(e) of the cGMP regulations states that:

*The accuracy, sensitivity, specificity, and reproducibility of test methods employed by the firm shall be established and documented. Such validation and documentation may be accomplished in accordance with § 211.194(a)(2).*

Section 211.194(a)(2) provides that:

> (a) Laboratory records shall include complete data derived from all tests necessary to assure compliance with established specifications and standards, including examinations and assays, as follows:
>
> (2) A statement of each method used in the testing of the sample. The statement shall indicate the location of data that establish that the methods used in the testing of the sample meet proper standards of accuracy and reliability as applied to the product tested. (If the method employed is in the current revision of the United States Pharmacopeia, . . . or in other recognized standard references, or is detailed in an approved new drug application and the referenced method is not modified, a statement indicating the method and reference will suffice.) The suitability of all testing methods used shall be verified under actual conditions of use.

Prior to trial, FDA criticized Barr's analytical method validation studies for a variety of reasons [43]. At trial, however, neither of the government's witnesses claimed to have any expertise in analytical methods or analytical methods validation [44]. Moreover, the only evidence submitted by the government from an analytical chemist was a written declaration that did not criticize any of Barr's methods [44].

The court found that an analytical method can be deemed validated if: (1) the method was approved as part of a firm's abbreviated new drug application (or, consequently, for a brand-name company, its new drug application); (2) the method is identical to the current USP method[3]; or (3) a firm conducts a satisfactory validation study [45]. The court also found that if a firm uses a method not recognized by USP or modifies a USP method, the firm must validate the method [45].

Because Barr's methods conformed to the USP or the CFR, or were approved as part of Barr's ANDAs, the court determined that the methods were entitled to a "*presumption of validity*" [45]. To the extent that Barr and the government disagreed about certain aspects of Barr's validation studies, the Court advised the parties to resolve

---

[3] Or, for antibiotic products, the applicable method set forth in the Code of Federal Regulations (CFR).

their differences informally, stating "*the Court is confident that the parties will be able to resolve any subsequent disagreements without judicial intervention*" [45].

## 4.4 Impact on the Industry

The Barr decision is disturbing for several reasons. First, it is not clear who is bound by the decision. Technically, only Barr and FDA in its dealings with Barr are required to adhere to the standards set by the court. In other words, if FDA were to sue another company for alleged cGMP violations, FDA could not automatically force the company to comply with the Barr standards (or vice-versa). Moreover, the decision has no binding effect on other cases. Even if the lawsuit were filed in the District of New Jersey, another judge would be entitled to reject or modify the standards.

Despite this, FDA seems determined to impose the Barr standards on other companies. FDA announced soon after the court issued its opinion that FDA investigators would carry the Barr decision with them when inspecting companies [46]. Since then, FDA has formally adopted the court's decision by issuing guidelines that essentially are identical to the opinion[4]. Thus, it appears that standards created by a judge, not scientists, with input from only one company and two FDA district offices, may govern proper laboratory and manufacturing practices for the entire industry.

Why has FDA chosen to set standards for cGMP through litigation? Why not regulate through notice and comment rulemaking like other agencies? FDA claims that "*[r]ulemaking takes time and resources. The FDA cannot always wait for rules to be in place before taking action*" [47]. As demonstrated below, however, there are good reasons why FDA should establish industry standards through notice and comment rulemaking.

The federal Administrative Procedure Act (APA) defines a "*rule*" as "*an agency statement of general or particular applicability and future effect designed to implement, interpret, or prescribe law or*

---

[4] In July 1993, FDA's Mid-Atlantic region formally adopted the Court's decision by issuing its "Guide to Inspections of Quality Control Laboratories." With few exceptions, the Guide mirrors the opinion.

*policy*" [48]. "`*Rulemaking' means agency process for formulating, amending, or repealing a rule*" [49].

In order to promulgate a rule, an agency must publish in the Federal Register a general notice of proposed rulemaking and must allow interested parties to participate in the rulemaking procedure [50]. The agency then must consider the comments submitted by the public and must delay the effectiveness of any final rule for at least thirty days [50]. Nothing in the Food, Drug, and Cosmetic Act suggests an exemption from the normal APA notice-and-comment procedures for FDA rulemaking.

The Administrative Conference of the United States has recommended that agencies employ notice-and-comment rulemaking when establishing policies that are to be binding or dispositive of issues [51]. The Administrative Conference recommendation dealt primarily with the issuance of policy statements, but the same danger exists when agencies make rules of general applicability through adjudication. The Administrative Conference comments are instructive:

> *The practical consequence is that this process may be costly and protracted, and that affected parties have neither the opportunity to participate in the process of policy development nor a realistic opportunity to challenge the policy when applied within the agency or on judicial review. The public is therefore denied the opportunity to comment and the agency is denied the educative value of any facts and arguments the party may have tendered.*

The Barr case illustrates the Administrative Conference's point. The standards created by Judge Wolin will affect an industry that had no opportunity to challenge, or even to be heard in, the process that led to the new rules.

The greatest problem with the Barr standards is that, although seemingly plausible on paper, they create unnecessary confusion in the laboratory. For example, although the court accepted the use of outlier tests to discard microbiological assay results, the Court refused to allow outlier testing for chemical assays. The Court determined that because USP did not discuss outliers in chemical testing, USP did not contemplate their use [17]. The Court stated that "[*t]he substantial variability of microbiological assays supports this distinction. Chemical assays are considerably more precise than biological and microbiological assays, since only the latter testing is subject to whims of microorganisms* " [17].

First, as Barr demonstrated at trial, the application of outlier tests to chemical assay data is acceptable and standard throughout the industry [52]. Second, as Barr's expert statistician explained in a subsequent commentary on the Barr case, the validity of outlier testing does not depend on variability because variability is taken into account in the test. Thus, an assay with large variability, such as a microbiological assay, would have to show considerable divergence due to the suspected outlier for the value to be rejected. Because of lower variability, testing for an outlier in a chemical assay might reject a less distant observation [53].

Third, nothing in the USP suggests that silence on a point is prohibitory. In fact, the USP implies that the use of scientifically sound statistical principles is acceptable:

> *Confusion of compendia standards with release tests and statistical sampling plans occasionally occurs. Compendial standards define what an acceptable article is and give test procedures that demonstrate that the article is in compliance. The manufacturer's release specifications, and compliance with good manufacturing practices generally, are developed and followed to assure that the article will indeed comply with compendial standards until its expiration date.*
>
> * * *
>
> *[USP tests on multiple dosage units] should not be confused with statistical sampling plans * * * Repeats, replicates, or extrapolations of results to larger populations are neither specified* <u>*nor proscribed*</u> *by the USP] compendia; such decisions are dependent on the objectives of the testing. Commercial or regulatory compliance testing, or manufacturer's release testing, may or may not require examination of additional specimens, in accordance with predetermined guidelines or sampling strategies* [54].

Even FDA's statistical expert[5] admitted that a firm may distinguish between an out-of-specification result caused by the "*random variability inherent in the data*" and an out-of-specification result that is the "*result of gross deviation from prescribed experimental procedure*" [55]. Where there has been a gross deviation, FDA's expert stated, a statistical outlier test can be a useful tool in a laboratory investigation [55].

Soon after the *Barr* decision, the USP Committee of Revision proposed a change to the General Notices section of the USP to indicate that the USP neither supports nor prohibits the use of outlier testing [56][6]. The USP adopted the proposed change to the General Notices section in USP 23, which became effective January 1, 1995 [57]. Thus, almost two years elapsed before the outlier testing issue was clarified by USP.

A second source of confusion is the court's ruling on retests. As discussed above, the Court stated that firms must do "*enough*" testing to reach a conclusion to confirm or reject an out-of-specification result. The Court then surmised that if a laboratory obtained 7 out of 8 passing results, that might be sufficient. Neither party proffered evidence indicating that 7 out of 8 testing would be an acceptable practice. Essentially, the Court pulled the 7 of 8 formula out of a hat[7] [58]. Now firms are left to wonder whether 7 of 8 is the

---

[5] FDA presented no live expert testimony on this issue at trial. In order to rebut Barr's testimony, FDA submitted a post-trial declaration from Dr. Thomas Hammerstrom, a mathematical statistician in the Biometrics Division of FDA's Center for Drug Evaluation and Research.

[6] "Repeats, replicates, statistical rejection of outliers, or extrapolations of results to larger populations are neither specified nor proscribed by the compendia; such decisions are dependent on the objectives of the testing." The General Notices section applies to all standards, tests, and assays set forth in the USP. Moreover, if a specific section of USP is silent on a particular issue, the General Notices section applies.

[7] In footnote 9 to the Opinion, the Court indicates that it relied on Compliance Officer Mulligan's testimony to arrive at the 7 out of 8 standard. Mr. Mulligan's testimony hardly supports the Court's conclusion:

Q. If you have one value in and one value out, how would you propose to proceed to determine what the true value is?

A. You would do additional testing on that same bottle.

Q. And how -- if you got two more passing results, you now have three passing and one out of specification results, would you then determine the bottle was within specification?

standard to which they must adhere. Moreover, the 7 of 8 formula raises questions about how much testing is enough, how much is too much, and when the amount of testing crosses the imaginary line and could prompt a claim that the firm is "*testing products into compliance*".

A third problem created by the Court is its position on averaging. As explained above, the Court determined that averaging might mask variability among initial values where some are within specification and some are not. Although the Court's point is well taken – out of specification data cannot be eliminated through averaging – the Court's solution is extreme. Just as out of specification results give a firm useful information about a batch, an average may give the firm more information about the true value of the batch than will individual results [59]. Moreover, because averaging is prohibited by the *Barr* Opinion, firms must devise new release criteria that may or may not be acceptable to FDA.

Another problem created by the court's ruling is that it provides scant information with respect to validation of methods. The decision states that analytical method validation is required (subject to the exceptions specified in the cGMP regulation), but it says nothing about how to structure validation studies that will comply with cGMP. The Court was unable to provide guidance of this nature because, as discussed above, the government did not present any evidence at trial to indicate what cGMP requires.

---

A. It's not -- I don't see that as conclusive.
Q. And if you take two more and you have five in spec and one out of spec, would that be conclusive to you, Mr. Mulligan?
A. No, but it's getting there.
Q. And if you took two more and had seven in spec and one out of spec, would that be conclusive to you?
A. I would defer to someone more expert in antibiotics to make that decision.

## 4.5 Conclusions

Assuming that the standards articulated in *Barr* are, in FDA's view, binding on other firms, there appears to be one positive aspect of the Opinion. As discussed above, in determining what constitutes cGMP, the Court held that to the extent the regulations are vague, firms can look to scientific literature or use "*scientific judgment*" where appropriate. This seems to invite industry to take the initiative in determining what standards should apply. Thus, with respect to outliers, retests and averaging (and, for that matter, any other aspects of the Court's opinion that are unacceptable to industry), industry should seek agreement concerning appropriate cGMP. If industry fails to do so, it will be left with the *Barr* decision or, perhaps, any number of court decisions, dictating scientific behavior.

## References

1. United States v. Barr Laboratories, Inc., *812 Fed. Supp.* 458 (D.N.J. 1993)

2. Barr Findings of Fact and Conclusions of Law (Proposed) ¶¶ 42-48, at 97-99, (D.N.J.) (No. 92-1744) (1993)

3. Barr Findings of Fact and Conclusions of Law (Proposed) ¶¶ 5-6, at 3-4, (D.N.J.)(No. 92-1744)(1993)

4. United States v. Barr Laboratories, Inc. 812 *F.Supp.* 458, 464 (D.N.J. 1993)

5. 21 U.S.C. § 351(a)(2)(B)

6. Government's Proposed Findings of Fact and Conclusions of Law ¶ 8, at 3, (D.N.J.) (No. 92-1744) (1993) (emphasis added)

7. United States v. Barr Laboratories, Inc., *812 F.Supp.* 458, 465 (D.N.J. 1993)

8. Transcript at 132-33

9. Transcript at 1044

10. Transcript at 225

11. Barr Findings of Fact and Conclusions of Law (Proposed) ¶¶ 49, 51, at 22-24, (D.N.J.) (No. 92-1744) (1993)

12. United States v. Barr Laboratories, Inc., *812 F.Supp.* 458, 466 (D.N.J. 1993)

13. United States v. Barr Laboratories, Inc., *812 F.Supp.* 458, 467 (D.N.J. 1993)

14. Barr Findings of Fact and Conclusions of Law (Proposed) ¶ 51, at 23, (D.N.J.) (No. 92-1744) (1993)

15. 21 C.F.R. § 211.165(d) (emphasis added)

16. Government's Proposed Findings of Fact and Conclusions of Law ¶ 35, at 15-16, (D.N.J.) (No. 92-1744) (1993)

17. United States v. Barr Laboratories, Inc., *812 F.Supp.* 458, 469 (D.N.J. 1993)

18. Transcript at 2192-94

19. Government's Proposed Findings of Fact and Conclusions of Law ¶ 31, at 12, (D.N.J.) (No. 92-1744) (1993)

20. Government's Proposed Findings of Fact and Conclusions of Law ¶¶ 17-27, at 7-10, (D.N.J.) (No. 92-1744) (1993)

21. Transcript at 1022

22. Transcript at 1533-34

23. Transcript at 57

24. Transcript at 60, 65-66

25. Government's Findings of Fact and Conclusions of Law ¶ 13, at 5-6, (D.N.J.) (No. 92-1744) (1993)

26. Barr Findings of Fact and Conclusions of Law (Proposed) ¶ 45, at 20, (D.N.J.) (No. 92-1744) (1993)

27. Barr Findings of Fact and Conclusions of Law (Proposed) ¶ 46, at 21, (D.N.J.) (No. 92-1744) (1993)

28. United States v. Barr Laboratories, Inc., *812 F.Supp.* 458, 470 (D.N.J. 1993)

29. United States v. Barr Laboratories, Inc., *812 F.Supp.* 458, 471 (D.N.J. 1993)

30. Government's Proposed Findings of Fact and Conclusions of Law ¶ 65, at 31, (D.N.J.) (No. 92-1744) (1993)

31. Government's Proposed Findings of Fact and Conclusions of Law ¶ 59, at 28, (D.N.J.) (No. 92-1744) (1993)

32 Transcript at 1558

33. Transcript at 1524-36

34. Barr Findings of Fact and Conclusions of Law (Proposed) ¶ 54, at 24-25, (D.N.J.) (No. 92-1744) (1993)

35. Transcript at 1527

36. Transcript at 1528

37. Transcript at 1529

38. Barr Findings of Fact and Conclusions of Law (Proposed) ¶ 56, at 25, (D.N.J.) (No. 92-1744) (1993)

39. United States v. Barr Laboratories, Inc., *812 F.Supp.* 458, 478-79 (D.N.J. 1993)

40. Transcript at 1747

41. Transcript at 1759

42. Transcript at 1763-65

43. Barr Findings of Fact and Conclusions of Law (Proposed) ¶ 7 at 4, (D.N.J.) (No. 92-1744) (1993)

44. Barr Findings of Fact and Conclusions of Law (Proposed) ¶ 8 at 4, (D.N.J.) (No. 92-1744) (1993)

45. United States v. Barr Laboratories, Inc., *812 F.Supp.* 458, 482 (D.N.J. 1993)

46. FDA Inspectors Will Carry Judge Wolin's Decision, Dickinson's FDA (Mar. 15, 1993)

47. M.J. Porter, Remarks by the Chief Counsel of the Food and Drug Administration: Enforcement, *Food and Drug Law J.* **47**, 143, 144 (1992)

48. 5 U.S.C. § 551(4)

49. 5 U.S.C. § 551(5)

50. 5 U.S.C. § 553

51. 1 C.F.R. § 305.92-2

52. See Transcript at 1394-95

53. S. Bolton, Should a Single Unexplained Failing Assay be Reason to Reject a Batch? *Clin. Res. Reg. Aff.*, **10**, 159, 174 (1993)

54. USP at 9 (emphasis added)

55. Hammerstrom Dec. at Sec. 11

56. *Pharmacopeial Forum*, **19**, (3), 5179, 5193 (1993)

58. Transcript at 804-05

59. See reply to Plaintiff's Response to Barr Laboratories' Motion for Clarification and/or Recommendation at 6

# Chapter 5

# Judge Wolin's Interpretations of Current Good Manufacturing Practice Issues Contained in the Court's Ruling in the United States v. Barr Laboratories

*Richard J. Davis*

## 5.1 Introduction

The decision handed down by Judge Wolin in the case of the United States *versus* Barr Laboratories [1] has had a profound effect on the way the pharmaceutical industry and the regulatory authorities view current Good Manufacturing Practices (cGMPs) [2]. Several similar injunctions had been issued previously against US pharmaceutical manufacturers, but they had all been settled out of court. Thus the "Barr case" was pivotal because it represents the first case of its kind in the United States that went to trial. This chapter provides an analysis of some of the potential implications of the judge's decision. The text has been adapted from a document written by this author and distributed as a guideline to various field offices of the Food and Drug Administration (FDA)[1] . Other views on this subject are to be found in Chapter 4 as well as in a recent review by Masden [3].

## 5.2 United States Pharmacopeia

There was much discussion at trial about USP standards and methods. The court ruled that USP's established standards are absolute and that firms cannot stretch the USP standards. These standards provide established criteria upon which firms release their product.

---

[1] During the "Barr Case", this author was Director of the Mid-Atlantic Regional Office of the Food and Drug Administration, Philadelphia, PA

## 5.3 Failure or out-of-specification results

Judge Wolin preferred to use the term "out-of-specification" (OOS) laboratory result rather than the term "product failure" which is more common to FDA's investigators. He ruled that an OOS result identified as a laboratory error by a failure investigation or an outlier test[2], or overcome by retesting[3] is not a product failure. OOS results fall into three categories:

- laboratory error
- non-process related or operator error
- process related or manufacturing process error

### 5.3.1 Laboratory errors

Laboratory errors occur when analysts make mistakes in following the method of analysis, use incorrect standards, and/or simply miscalculate the data. Judge Wolin provided specific guidance on the matter of determining when an error can be designated a laboratory error. Laboratory errors must be determined through a failure investigation to identify the cause of the OOS. Once the nature of the OOS result has been identified, it can be classified into one of the three categories above. He states that the inquiry may vary with the object under investigation.

### 5.3.2 Laboratory investigations

The court said that the exact cause of analyst error or mistake can be difficult to pin down and that it is unrealistic to expect that analyst error will always be determined and documented, and he ruled that the "laboratory investigation consists of more than a retest". The inability to identify the cause an error with confidence affects retesting procedures, not the investigation inquiry required for the initial OOS result.

---

[2] The court provided explicit limitations on the use of outlier tests and these are discussed later in this chapter.
[3] The court ruled on the use of retesting, which is covered later in this chapter.

The analyst should follow a written procedure, checking off each step as it is completed during the analytical procedure. Laboratory test data must be recorded in notebooks; use of scrap paper and loose paper is to be avoided. These measures enhance the investigation process. The court specifically identified procedures that must be followed when **single** and **multiple** OOS results are investigated.

For the **single** OOS result, the investigation must include the following steps and these inquiries must be conducted before there is a retest of the sample:

a) the analyst conducting the test must report the OOS result to the supervisor

b) the analyst **and** the supervisor must conduct an informal laboratory inspection which addresses the following areas:

- Discuss the testing procedure
- Discuss the calculation
- Examine the instruments
- Review the notebooks containing the OOS result

An alternative means to invalidate an initial OOS result provided the failure investigation proves inconclusive is an "outlier" test. The Court placed specific restrictions on the use of such a test.

a) Firms cannot frequently reject results on this basis

b) The USP standards govern its use in specific cases

c) An outlier test cannot be used for chemical testing results[4]

---

[4] An initial content uniformity test was OOS followed by a passing retest. The initial OOS result was claimed the result of analyst error based on a statistical evaluation of the data. The use of an outlier test is inappropriate in this case.

A full scale inquiry is required for **multiple** OOS results. This inquiry involves quality control and quality assurance personnel **in addition to laboratory workers** to identify exact process or non-process related errors. The court ruled that when the laboratory investigation is inconclusive (reason for the error is not identified), the laboratory:

a) cannot conduct two retests and base release on average of three tests

b) cannot use outlier test in chemical tests

c) cannot use a resample to assume a sampling or preparation error

d) is allowed to conduct a retest of different tablets from the same sample when a retest is considered appropriate (see criteria elsewhere)

### 5.3.3 Formal Investigations

Judge Wolin ruled that formal investigations extending beyond the laboratory must follow the government's outline with particular attention to corrective action. He said the company must:

a) State the reason for the investigation

b) Provide summation of the process sequences that may have caused the problem

c) Outline corrective actions necessary to save the batch and prevent similar recurrence

d) List other batches and products possibly affected, the results of investigation of these batches and products, and any corrective action. Specifically:

- examine other batches of product made by the troublesome employee or instrument

- examine other products produced by the troublesome process or operation

e) Preserve the comment and signature of all production and quality control personnel who conducted the investigation and approved any reprocessed material after additional testing

### 5.3.4 Investigation documentation

Analysts' mistakes, such as calculation errors, should be specified with particularity and supported by evidence. Investigations along with conclusions reached must be preserved with written documentation that enumerates each step of the review in the form of a "computer generated flow sheet". This writing should be preserved in an investigation or failure report and placed into a central file.

### 5.3.5 Investigation time-frames

All failure investigations must be performed within 30 business days of the problem's occurrence and recorded and written into a "failure or investigation report".

### 5.3.5 Product failures

An OOS laboratory result can be overcome (i.e. disregarded) when laboratory error has been documented. However, non-process and process related errors resulting from operators making mistakes, equipment (other than laboratory equipment) malfunctions, or a manufacturing process that is fundamentally deficient, such as an improper mixing time, represent product failures.

## 5.4 Retesting

Several opinions about retesting were issued in this decision. The number of retests performed before a firm concludes that an unexplained OOS result is invalid or that a product is unacceptable is a matter of scientific judgment. The goal of retesting is to isolate OOS results but retesting cannot continue *ad infinitum*. In the case of non-process and process-related errors, retesting is suspect. Because the initial tests are genuine, in these circumstances, additional testing alone cannot infuse the product with quality. The court acknowledges that some retesting may precede a finding of non-process or process-based errors. Once this determination is made, however, additional retesting for purposes of testing a product into compliance is not acceptable.

For example, in the case of content uniformity testing designed to detect variability in the blend or tablets, failing and non-failing results are not inherently inconsistent and passing results on limited retesting do not rule out the possibility that the batch is not uniform. As part of the investigation, firms should consider the record of previous batches, since similar or related failures on different batches would be a cause of concern.

**A very important ruling in this decision sets forth a procedure to govern the retesting program.** The judge ruled that a firm should have a predetermined testing procedure and it should consider a point at which testing ends and the product is evaluated. If results are not satisfactory, the product is rejected. Additionally, the company should consider all retest results in the context of the overall record of the product. This includes the history of the product[5] type of test performed, and in-process test results. Failing assay results cannot be disregarded simply on the basis of the content uniformity results being acceptable.

Retesting following an OOS result is ruled appropriate only after the failure investigation is underway and the failure investigation determines, in part, whether retesting is appropriate. It is appropriate when analyst error is documented or the review of analyst's work is "inconclusive", but it is not appropriate for non-process or process-related errors.

---

[5] The court ordered a recall of one batch of product on the basis of an initial content uniformity failure and no basis to invalidate the test result and on a history of content uniformity problems with the product.

The court ruled that retesting:

- must be done on the same, not a different sample
- may be done on a second aliquot from the same portion of the sample that was the source of the first aliquot
- may be done on a portion of the same larger sample previously collected for laboratory purposes

## 5.5 Resampling

Firms cannot rely on resampling[6] to release a product that has failed testing and retesting unless resampling is in accord with the USP standards (content uniformity and dissolution), or unless the failure investigation discloses evidence that the original sample is not representative or was improperly prepared.

## 5.6 Averaging results

Averaging can be a rational and valid approach, but as a general rule this practice should be avoided[7] because averages hide the variability among individual test results. This phenomenon is particularly troubling if testing generates both OOS and passing individual results which when averaged are within specification. Here, relying on the average figure without examining and explaining the individual OOS results is highly misleading and unacceptable.

Content uniformity results never should be averaged to obtain a passing value for content uniformity.

In the case of microbiological assays an average is preferred by the USP. Also, the Judge ruled that it is good practice to include

[6] The court ordered the recall of one batch of product after having concluded that a successful resample result alone cannot invalidate an initial OOS result.

[7] The court ruled that the firm must recall a batch that was released for content uniformity on the basis of averaged test results.

OOS results in the average, unless an outlier test (microbiological assays) suggests the OOS is an anomaly.

### 5.7 Remixing

The need to remix often is a clear indication that the process is invalid and casts doubt on those batches passed through testing without incident. Remixing is reworking permitted under the GMP regulations. Occasional remixing is acceptable, but frequent or wholesale remixing is unacceptable.

### 5.8 Product release

Scientific judgment can play a role when firms decide to release a batch to the public and the court said it cannot articulate specific procedures for release decision making. However, Judge Wolin said that the USP standards upon which firms release their products are absolute and cannot be stretched. For example, a limit of 90 to 110 percent of declared active ingredient, and test results of 89, 90, 91, or two 89s and two 92s should be followed by more testing.

It is clear that the release evaluation depends in part on the background of the batch and product. Secondary factors that affect the actual finished product results as well as their reliability are:

- Physical properties
- Blend evaluations
- Time of mix
- Tablet weight, thickness, and friability

Judge Wolin ruled that context and history[8] inform many final conclusions and that one must consider past problems with the

[8] On the basis of an initial content uniformity failure and no basis to invalidate the test result and on a history of content uniformity problems with the product, the court ordered the batch recalled.

product and batch and evaluate all the data relative to the product and batch.

## 5.9 Blend testing

Blend testing is necessary to increase the likelihood of detecting inferior batches. Blend content uniformity testing cannot be waived in favor of total reliance on finished product testing because finished product testing is limited.

The court ruled that sample size influences ultimate blend test results and that the sample size should resemble the dosage size. Any other practice would blur differences in portions of the blend and defeat the object of the test. **The appropriate sample size for blend content uniformity in both validation and regular production batches is three times the active ingredient dosage size.**

Multiple individual samples taken from different areas cannot be composited. However, when variation testing is not the object of assay testing, compositing is permitted.

Firms must demonstrate through validation that their sampling technique is representative of all portions and concentrations of the blend. This means that the samples must be taken from places that might be problems, weak or hot spots in the blend.

In this case, the firm maintained that samples could be collected from the drums containing the finished blend. The court ruled that the firm must demonstrate that sampling from drums rather than the mixer is representative. He also ruled that the firm cannot composite blend samples and that they must take smaller blend content uniformity samples.

## 5.10 Validation

### 5.10.1 A retrospective validation

The court ruled that batches meeting the following criteria must be included in retrospective validation studies:

a) All batches made in the time period specified for study must be included unless the batch was made from a non-process related error

b) Only batches made in accord with the process being evaluated can be included

Only test results determined through an appropriate failure investigation and found to be caused by analyst or operator error can be excluded from the study. Test results that are explained but merely called into question by successful retesting must be included in the study. **The exclusion of batches and test results must be documented through failure investigation.**

The number of retrospective batches chosen for the study must be greater than the number used for prospective validation. Although the court set no exact number of batches to be chosen, guidelines have been established as follows:

a) Five batches is unacceptable and also 6 or more may not be acceptable

b) Because a 10% batch failure is unacceptable, if one batch fails, more than 10 batches are needed for the retrospective study

c) Experts accept 20 to 30 batches

### 5.10.2 Concurrent and prospective validation

Concurrent and prospective validation requires at least three consecutive batch runs of the process. **Mixing time** studies should be included in a prospective validation program and follow any problems that surface in retrospective validation batches. **Particle size distribution specifications** are widely accepted industry practice and should be included in validation studies.

## 5.11 Significance of application approvals

The court ruled that a firm cannot rely on the claim that the FDA previously approved their procedures contained in an approved application. This approval cannot be used as a defense and cannot be used to shield a process that produces failures.

## 5.12 Methods validation

Methods can be validated in a number of ways. Methods appearing in the USP are considered validated and they are considered validated if part of an approved ANDA. Also, a company can conduct a validation study on their method. System suitability data alone is insufficient for method validation.

## 5.13 Cleaning validation

The court ruled that a firm cannot wait for contamination and other problems to reveal inadequate cleaning procedures. In order for the cleaning rules to be effective, the specific methods chosen must be shown to be effective. The court ruled that a milling machine is a major piece of equipment and must be included in the cleaning validation program. Firms must identify the cleaning agents used in its cleaning process. When these agents are known to cause residue, the company must check for the residue provided the firm has described its cleaning methods and materials in sufficient detail **and** unless the cleaning material is known to cause a residue **then** one run through of the cleaning procedure, in the absence of problems, is not insufficient for validation.

## 5.13 Conclusions

Judge Wolin rendered a decision that is consistent with the concept of building quality into the product. He relied heavily on the opinions of the experts presented by both sides. The judge's decision instructs both the government and the industry in the doctrines and concepts that must be considered when evaluating difficult scientific issues affecting the quality of drug products. This is most evident in the discussions of procedures for the investigation of out of specification results and the factors one must address when considering the release of drug products.

This decision concludes that end product test results represent only one of many indications of quality and these results alone represent an insufficient quality release specification.

## References

1. 812 *Federal Supplement* 458 (D.N.J. 1993)

2. 21 CFR §210.1-3, 211.1-208

3. R.E. Masden, *PDA J. Pharm. Sci. Technol.*, **48**, 176-179 (1994)

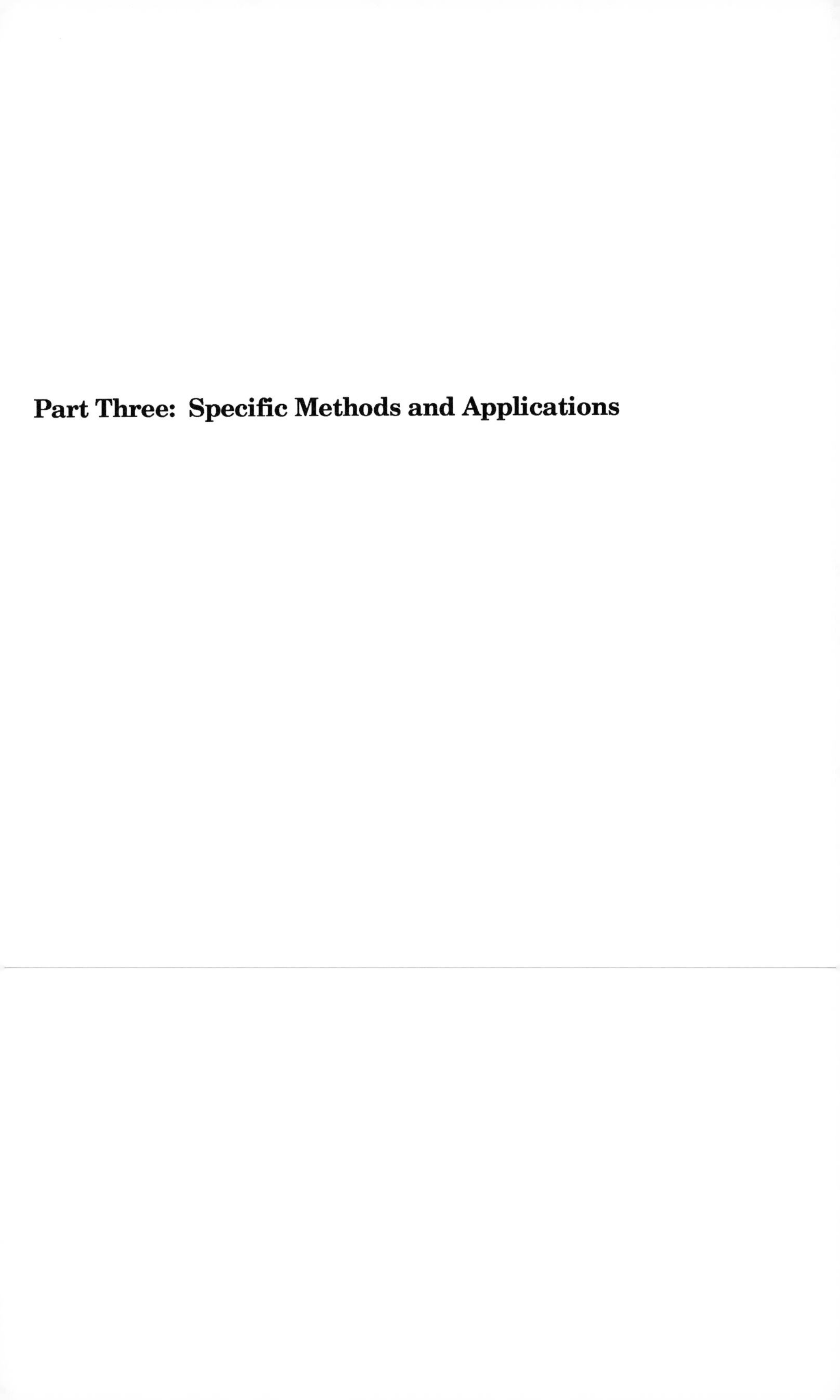

# Part Three: Specific Methods and Applications

## Chapter 6

# Bulk Drug Substances and Finished Products

*Paul K. Hovsepian*

## 6.1 Introduction

Any pharmaceutical formulation, used in clinical investigations or approved for marketing, must meet regulatory requirements of identity, quality, strength and purity. In the development of a new dosage form, all starting materials, synthetic intermediates, drug substances, excipients and formulated products must be tested for compliance with the sponsor's in-house and regulatory acceptance criteria. Such testing must include quantitation of synthetic impurities and degradation products including their isolation and characterization, where appropriate. In addition, analytical controls must be established to demonstrate that the synthetic or manufacturing process is capable of producing the formulated product in a consistent and reproducible manner. The United States Food and Drug Administration recognizes (21 CFR 312.23) that drug development[1] is an evolving process with ongoing changes, as the investigation progresses in the synthesis of the drug substance and formulation of the dosage form. Thus, the technical information submitted by the sponsor for an Investigational New Drug (IND) is focused more on the identity and control of the raw materials, the drug substance and formulation ensuring they are within acceptable chemical and physical limits for the planned duration of the proposed investigation. The sponsor is therefore responsible for submitting additional information to amend the IND application as appropriate, consistent with the phase of chemical development.

This chapter will review the analytical test methods and controls which are commonly used to establish the quality of bulk

[1]This chapter was developed within the United States' Regulatory framework. However, the material is generally applicable to regulations in other countries.

pharmaceuticals and pharmaceutical formulations with special emphasis on how these methods evolve during drug development phase. The primary focus will be on chromatographic test methods for assays and related substances.

## 6.2 Bulk Drug Substance

Even though the GMP regulations, under 21 CFR parts 210 and 211 apply only to finished dosage forms, bulk drug substances are manufactured in accordance with concepts of current good manufacturing practice (cGMPs) since these are components of drug products and, hence, not exempt from complying with cGMP principles. In practice, it is very difficult and often unnecessary to apply rigid controls to bulk drug manufacture during the early processing steps of the synthesis. However, the controls which may vary from synthesis to synthesis, must become rigid in the final steps of the synthesis.

### 6.2.1 Raw Materials

Analytical control of raw materials is required to minimize impurity levels and to ensure the quality of the end product. This typically includes an identity test, a chromatographic assay (i.e. GC or LC) and compendial testing if the reagents, solvents and/or reactants are of compendial grade. It is difficult to set rigid specifications at the early stages of the synthetic process. Usually target limits are assigned as controls, and coupled with chromatographic profiles, they provide adequate screening of the early materials. It is imperative that the test procedures and target limits be written to ensure they are applied routinely and that they are well documented.

### 6.2.2 Starting Materials

These are generally commercially available chemicals, which are incorporated in the new drug substance as important structural elements. Generally, the decisions about what comprises the starting material is by agreement between the sponsor and the FDA chemist, usually at a late Phase 2 or early Phase 3 meeting. Data for the starting material are not part of an IND submission but could be submitted as an amendment as they become available. However, it is a requirement to provide data on the identity, assay and impurities of a starting material as part of the NDA submission. Starting materials must be controlled, especially in those cases where impurities are present and could be carried through to the drug substance. Usually, an identification test, LC or GC assay and impurity measurement by TLC or LC are sufficient to provide analytical characterization of the starting material. The test methods and controls must be documented to ensure their routine application.

### 6.2.3 In-process Controls

At selected intermediate stages of the chemical synthesis, test methods and controls are necessary to ensure that the synthetic process is under control. Early in development, almost every step of the synthesis is monitored, usually by a chromatographic or spectroscopic method which provides an understanding of the process in addition to providing an estimate of yield and purity. In-process controls become a routine part of the synthesis by Phase 3. At this point, specifications for intermediates are in place and test procedures are well established, meeting the requirements of the regulations and becoming part of the process validation of the chemical synthesis. The analytical controls should include a key physical property such as optical rotation, refractive index, melting point in addition to the assay and impurity measurements. This information is used to determine the yield during the early stages of the synthesis and ensures that the system is operating within the expected range. Specifications for the final intermediate should be almost as tight as those of the drug substance since this is the last synthetic step in which the purity is monitored before converting to the drug substance. Batch data on the final intermediate (penultimate) are normally not submitted at an IND, however the sponsor must have a formal process for testing and control of the final intermediate if it is isolated, since this information is required for the NDA submission. Finally, the analytical test methods and controls for the final intermediate should include an identification such as IR, NMR, a physical property such as, optical rotation, an assay such as GC, LC, or titration and a method for impurities.

### 6.2.4 Release Tests for Bulk Drug Substance

Analytical test methods and specifications for bulk drug substances must be developed by the time of IND filing to ensure the regulatory agencies that the proper identity, quality, strength and purity requirements have been achieved in the synthesis. The following physical and chemical attributes are measured with each drug substance with appropriate specifications to ensure the synthetic process has successfully produced a chemical entity with the requisite properties for therapeutic activity.

#### 6.2.4.1 Description/Appearance

Description is submitted in a narrative form describing the color and crystalline form of the drug substance.

#### 6.2.4.2 Physical Properties

Examples of physical properties could include melting point, boiling point, refractive index, viscosity, optical rotation, particle size, solubility, bulk density or volume and polymorphism.

#### 6.2.4.3 Identity Tests

Identity tests must be specific and capable of differentiating the drug substance from contaminants. The US Food and Drug Administration recognizes the following as primary identity tests, IR/FTIR, nuclear magnetic resonance (NMR), and mass spectrometry (MS). Identity tests such as UV spectra, or chromatographic attributes such as $R_f$, $t_r$ are considered confirmatory techniques but are not substitutes for specific identity tests.

#### 6.2.4.4 Assays

A precise and accurate quantitative procedure is required to determine the purity of the drug substance, especially since the control limits are usually very tight (usually 98.0-102.0% on a solvent free basis). The most commonly used technique for pharmaceutical assays is liquid chromatography because of its simplicity, good specificity, sensitivity and excellent precision and accuracy. Specificity is highly desirable to ensure the complete separation of the analyte from related substances for stability indicating purposes. However, if method specificity is not achievable, a precise non-specific quantitative assay such as titrimetry, spectrophotometry, coulometry or differential scanning calorimetry can be used coupled with a specific assay for the related substances.

### 6.2.4.5 Impurities

According to the USP <1086> "An impurity is any component of a drug substance (excluding water) that is not the chemical entity defined as the drug substance".

Impurities could be related substances i.e. structurally related to the drug substance such as degradation products or they can arise from the synthetic process as process contaminants such as side products, catalysts, reagents or heavy metals, or could be residual organic solvents, i.e. crystallization solvent. Measurement and control of impurities is important in product development. Tracking all impurities both known and unknown is intended to assure the safety of the final product.

Setting impurity limits in drug substances is an evolutionary process which begins before an IND is filed, to initiate clincial trials, and continuing well after the approval of a new drug application (NDA). It is the drug manufacturers responsibility to assure the regulatory agencies that the analytical methods used to evaluate impurities in a drug substance, during any portion of the development phase, are suitable for their intended purpose. The following process is representative of how impurity limits are established in the drug development process.

#### 6.2.4.5.1 IND, Phase 1 and 2

During this phase the synthesis is at an early stage of development and only a few batches of drug substance have been produced at a laboratory or semi-works scale. Limits are set on total impurities and an upper limit is set for any single impurity. However, for impurities known to be toxic, limits should be set no higher than the level that can be qualified (i.e. justified by safety data). Residual solvent limits are based on the known toxicology of the solvent and proposed dosing requirements.

Other process contaminants are monitored by the appropriate compendial methods and specific metal contaminants are controlled based on their toxicological properties.

#### 6.2.4.5.2 Phase 3 and NDA

At the time of filing the NDA there should be a large database of information on impurities in the drug substance manufactured by the process proposed for filing. Methods should be fully validated and results of toxicological studies should have qualified some of the related substances. Additionally all impurities $\geq$0.10% should have been isolated and characterized according to the final ICH guideline on impurities [1]. Based on the above information, the impurity limits filed in the original IND are revised and adjusted accordingly. Limits for process contaminants and residual solvents are set based on their toxicological properties. Also if appropriate, a method for enantiomeric purity is fully validated and the purity controlled.

#### 6.2.4.5.3 Post NDA

After the NDA is approved, there could be changes in the process for the manufacture of the bulk drug substance. The innovator is required to review the impurity profile of the drug substance to determine if the toxicological studies are still supportive and take the proper action. The impurity limits and rationale for setting them must be reviewed, justified and adjusted accordingly.

## 6.3 Drug Product

The goal of product manufacturing is to ensure each batch meets the identity, strength, quality and purity it is purported to possess.

### 6.3.1 Analytical Methods and Specifications

Laboratory controls should include adequate specifications and test methods to assure that formulation ingredients, in-process and finished products, conform to established standards of identity, strength, quality and purity. Typically these specifications are identical to, or more stringent than those contained in the compendia themselves.

The FDA recognizes that specifications will continue to evolve due to formulation and/or manufacturing changes. The same is true for the analytical methods, and often different methods are used in response to a formulation change causing an interference with the current test method. The extent of testing is dependent on the development phase, and as the project matures, methods become more entrenched and controls more specific. Thus, at an early stage of development acceptance/rejection criteria may not be as specific; however, such criteria must be based on sound scientific principles.

Analytical test methods and controls should include the following:

#### 6.3.1.1 Excipients

All excipients must meet compendial requirements. Tests which are not part of the compendia must be validated by the time of NDA filing.

#### 6.3.1.2 In-Process Controls

Analytical tests such as composite assay using LC for granulation, or UV for bulk solutions are commonly used to control in-process bulk formulations and appropriate adjustments can be made if necessary to bring the product with acceptable limits.

#### 6.3.1.3 Drug Product Testing

The following tests were selected to ensure quality products and are typical measurements required to ensure that the dosage form will meet the criteria for identity, quality, potency and purity. The list is by no means exhaustive and the selection of tests depends on the type of dosage form under evaluation. Those for the most common dosage forms are listed below.

a) Solid Dosage Forms (e.g. Tablets, Capsule)

- Identification (IR, NMR, etc.); also define color, and shape for tablets
- Moisture content
- Composite assay
- Degradation products
- Content uniformity per applicable compendia
- Dissolution per applicable compendia
- Disintegration
- Hardness and friability for tablets, friability for capsules

b) Solutions, Injectibles, Lyophilized Dosage Forms

- Identification

- pH of solution
- Clarity and particulate matter of injectibles and constituted solutions
- Sterility of injectibles
- Endotoxin levels per USP <85> or other applicable compendia
- Microbial limits test per USP <61> or other applicable compendia
- Antimicrobial preservative - effectiveness test per USP <51> or other applicable compendia
- Potency (assay)
- Degradation products

In the examples above, placebos are tested where appropriate to demonstrate the absence of the active drug substance.

### 6.3.1.4 Dosage Form Assay

With few exceptions, the stability indicating assay method developed for the drug substance is adapted for dosage form assay. If method specificity is not achievable, a precise non-specific quantitative assay method is used coupled with a specific assay for related substances. Regardless of the method used, the analyst has to make certain that samples selected are representative of the batch.

a) Solid Dosage Forms

Tablets and capsules are the most common types of pharmaceutical dosage forms. Liquid chromatography is the most common method for measuring the potency of pharmaceuticals because very large numbers of these compounds are polar organic compounds and are easily chromatographed by reversed-phase LC. The drug potency is measured versus a reference standard under conditions where matrix interference is absent.

When assaying solid dosage forms the largest source of quantitation error can be attributed to the sample preparation step, especially in the case where drug extraction from the sample matrix is required. Commonly, the drug product assay is controlled between 90% to 110% of the label. More recent product specifications, however, tend to be set between 95.0% to 105.0% as a result of worldwide harmonization of specifications.

b) Liquid Dosage Forms

Oral liquids and injectibles including lyophilized products are the most common liquid dosage forms. Assay measurements are usually carried out by LC. In addition, sterility, pyrogenicity and preservative assay as well as antimicrobial efficacy challenge tests are part of the acceptance criteria for these products. The latter will be described in more detail in this chapter.

c) Degradants in Liquid and Solid Dosage Forms

Liquid chromatography with reversed-phase gradient elution is typically used to separate degradants from the drug substance, synthetic impurities and excipients present in liquid or solid dosage forms. Quantitation of eluting components are usually carried out using UV detection at a selected wavelength. However, since UV detector response can vary greatly from compound to compound at a given wavelength, it is desirable to report results in terms of weight percent rather than area percent to avoid potential misleading results. This can be achieved by assaying the degradant versus a sample of pure authentic standard of the same compound as an external standard. However, it is not always practical to obtain large quantities of pure authentic compound for day-to-day analysis. For this reason, most methods employ relative response factors to quantitate the chromatographic peak of interest.

A relative response factor is calculated as the ratio of the molar absorptivity of the authentic standard (degradant or impurity) to the molar absorptivity of the reference standard of the active drug substance. Response factors are determined by chromatographing solutions of authentic standard and drug substance reference standard each at a concentration which yields approximately equal area counts.

The weight percent data calculated using response factors are accurate as long as the chromatographic parameters such as detector sensitivity, pH, ionic strength of the mobile phase, solvent strength, and UV wavelength used in the analysis of the sample are identical to the LC conditions used in developing the response factor [2].

## 6.4 Analytical Test Methods and Specifications for Drug Products

The analytical controls and test methods for drug products go through an evolutionary process, and as discussed below, the controls and test methods must be suitable for the intended purpose at each stage of development.

### 6.4.1 IND and Phases 1 and 2

Tests and specifications at this stage of development are generated to ensure the safety and stability of the dosage form during the clinical studies. Typically the tests include a primary identity test, an assay, degradants, uniformity of dosage unit and dissolution profile for solid dosage forms and appropriate microbiological tests for injectibles.

The specifications during IND and phase 1 and 2 are usually tentative since the product is not yet fully developed. The ultimate goal of this stage of product development is to ensure that the formulation is safe and stable during the duration of the clinical studies.

### 6.4.2 Phase 3

Methods and specifications are fully developed during this phase including any physical or chemical characteristics of the dosage form. Additionally the specifications are tightened based on batch analysis data.

### 6.4.3 Post NDA

After approval of the NDA any changes in the formulation are evaluated by comparing the new formulation with the biobatch to ensure bioequivalence. Additionally, analytical methods are adjusted if necessary and cross-validated against the NDA method for equivalency. The NDA is supplemented as appropriate.

## 6.5 Stability

Stability testing provides information on how the quality, safety and/or efficacy of a drug substance and product are affected by environmental factors such as temperature, humidity and light. The studies are designed to show that the drug substance and product will remain within established specifications during the re-test period if stored under the recommended storage conditions.

### 6.5.1 Drug Substance Stability

Any stability study must be conducted according to a written stability protocol designed to satisfy the developmental phase of the study. As per 21CFR312.23, data are required to support the stability of the drug substance during the pre-clinical toxicological studies, and in all phases of the IND to ensure that the drug substance is well within the acceptable chemical and physical limits for the planned duration of the proposed clinical investigation. The drug substance stability profile is usually developed early in the development process. The preliminary stability profile becomes very useful in the development of a stability protocol. It is important to note that there is no expiration dating required for a drug substance unless it is an antibiotic. The physiochemical changes of a drug substance on stability are quantitated using stability indicating chromatographic methods. These methods are used to determine the intrinsic stability of a new molecular entity by establishing degradation pathways to project the likely degradation products. Stability data from accelerated and long term testing is required for registration application. The laboratory scale drug substance batches placed on stability representing batches used in pre-clinical and clinical studies, are submitted as supportive stability data for registration.

Formal stability data are generated by placing batches of drug substance manufactured on a pilot scale using the final synthetic process. This is the formal stability data required in the new drug application. Post approval, three production size drug substance batches are placed on long term stability using the identical stability protocol as in the approved drug application.

Stability storage conditions for NDA registration should contain the following (USP <1196>):

- Long term testing 25° C ± 2° C/60% RH ± 5% with a minimum of 12 months data.
- Accelerated condition 40° C ± 2° C/75% RH ± 5% with a minimum of 6 months storage.

When a "significant change" occurs during 6 months storage under accelerated conditions as listed above, additional testing at an intermediate condition (such as 30° C ± 2° C and 60% RH ± 5%) should be conducted for drug substance to be used in the manufacture of dosage forms and tested long term at 25° C 60% RH. This information is included in the registration application.

### 6.5.2 Drug Product Stability

The stability protocol for a pharmaceutical product is designed using the stability properties of the drug substance. Data on early batches are used to support early clinical studies. Following the preliminary formulation and subsequent development of a stability-indicating assay, the formulator-analyst team designs and conducts a special study to demonstrate product stability. The formulation is stored at various stress and accelerated conditions (elevated temperature, light and humidity) to produce a stability profile in a relatively short period of time, thereby providing the needed assurance that the product will be stable for the intended duration of the clinical studies. As the project develops, data on associated formulations and packages are used as supportive stability data for NDA registration. Stability data, accelerated and long-term, on three batches of the same formulation and dosage form, including container and closure proposal for marketing, are submitted as primary data for NDA registration. It is expected that the above batches are at least pilot scale stored 12 months at room temperature (25° C, 60% RH) and 6 months at 40° C/75% RH in market image packages. The analytical tests should cover those features which are expected to change during storage. All products should be tested for their physical and chemical properties and where appropriate for their microbiological attribute.

The acceptance criterion for the product release is set tighter than for shelf-life to allow for loss of potency during storage. However, this must be justified based on data from several batches. In addition, any difference between the release and shelf-life specifications for antimicrobial preservatives should be supported by preservative efficacy testing.

## 6.6 Microbiological Attributes for Non-Sterile Drug Substances and Raw Materials [2]

The purpose of any microbiological control program for non-sterile pharmaceuticals is to locate and eliminate possible sources of contamination in order to manufacture a finished product of the quality appropriate for its intended use. There should be no requirements for bacterial endotoxin specification for excipients or drug substances intended for use in non-sterile products.

The need for microbiological control of non-sterile drug products that are susceptible to microbial contamination is well recognized by the pharmaceutical industry, regulatory agencies and the official compendia (USP-XXIII <1111>) (CFR21 211.84: 113: 165).

The primary goal of a microbiological testing program is to control the type and number of microorganisms to which the final product is exposed. This is accomplished by the control of microbial bioburden of raw materials, processing equipment and facilities, packaging components, and active drug substances. Few raw materials used in making pharmaceutical products are sterile when received. In general, purified or processed materials of synthetic origin exhibit low bioburden levels. Higher bioburden levels are usually seen in material derived from natural products. Raw materials, bulk in-process materials and drug substances must be maintained under storage conditions which minimize microbial proliferation and monitored periodically for bioburden levels.

Drug components can be a primary source of microbial contamination. Sampling plans, test procedures and specifications are developed for the control. Materials of animal or botanical origin require extensive testing to estimate the level and extent of bioburden, whereas synthetic materials require minimal testing since they usually exhibit low bioburden levels.

In the manufacture of non-sterile pharmaceutical products, effort is made to design a formulation which meets the desired microbiological attributes throughout the products claimed (labeled) shelf life. To achieve this, many products are preserved against the growth of microorganisms such as by pH adjustment, or by the addition of chemicals which are compatible with the formulation and have known preservative value.

Many non-sterile products provide an adequate medium for the growth of a broad spectrum of microorganisms. The presence of large number of certain microorganisms may reduce the quality of non-sterile pharmaceutical products as evidenced by decomposition, odor, change in flavors, discoloration and production of toxins when microbiological levels are not controlled. Therefore, for the protection of the consumer it is the manufacturer's responsibility to maintain low levels of microorganisms in their products.

Specifications should meet the requirements of the major compendia involved in the international harmonization effort. In general, test methods are based primarily on the USP. If alternate methods are developed and validated, they must be equivalent to or better than the USP method. When more than one compendium governs a test, the method will be based on the most stringent, yet scientifically sound, test requirements as determined by experimental data.

### 6.6.1 Excipients and Drug Substances

Excipients and drug substances that have inherent antimicrobial activity or materials such as chemicals with high or low pH generally do not need to be monitored for bioburden. Data should be generated to confirm the presence of antimicrobial activity by performing USP <61> Microbial Limits Tests.

Processed or purified synthetic materials generally have a low bioburden and may not need to be monitored on a regular basis. These materials may be partially antimicrobial or have no antimicrobial properties, which should be demonstrated by laboratory studies. For those materials that have no antimicrobial properties and are expected to contain bioburden based on the nature or origin of the material, routine testing should be performed on each lot. Historical data needs to be taken into consideration when establishing limits and specifications.

High bioburdens are typically seen in materials derived from natural sources. These materials should be tested on a lot by lot basis, and results compared against specifications. It may be necessary for materials of natural origin (plant, mineral, etc.) that typically have high bioburden to have higher specifications in comparison to those materials chemically synthesized that typically have low bioburden. (For example, tapioca starch 5000 cfu/g[2] total count vs. microcrystalline cellulose 500 cfu/g total count). Specifications must be based upon safe allowable limits for that material or product. There should be documented evidence that the higher bioburden has no deleterious effect on the finished product.

Indicator organisms selected for excipients and drug substances should be consistent with those selected for finished products containing those materials. Indicator organisms are not limited to the four USP <61> indicator organisms, but should be based upon formulation attributes, the nature or origin of the excipients, the potential impact on the user, route of administration and intended use. If bioburdens are consistently low, it may be possible to monitor the quality of a material of synthetic origin by total count alone.

When establishing microbial specifications, it is important to understand the manufacturing process of the suppliers, and their ability to provide the level of microbial control required. Consideration should be given to vendor's specifications and testing requirements when establishing in-house specifications for new excipients and drug substances. The supplier's operation should be in compliance with cGMPs and operating in a manner that effectively controls the possibility of microbial contamination during manufacture, sampling, testing, storage and shipment to the customer. These factors should be confirmed during a Quality

---

[2] cfu/g = colony forming units per gram

Assurance (QA) audit of the supplier. Consideration may be given to accepting vendor results if the vendor's testing facilities and methods have been audited and approved by the QA department.

Lower alert and action limits are considered for additional quality control, as necessary. Alert levels are microbiological quality levels or ranges which, when exceeded, indicate a potential drift from normal quality conditions. If exceeded, action may not be required, but may indicate a need to monitor the material quality more closely than regularly programmed. Alert limits may be implemented to allow for earlier detection of a trend and identification of potential future lots to be rejected. Results that exceed alert limits are documented in a manner that ensures notice and evaluation of a potential trend.

For some excipients and drug substances, it may be prudent to establish action limits that are tighter than the specifications. Action levels are microbiological quality levels or ranges which, when exceeded, indicate an apparent drift from the normal quality conditions and require immediate action. Lots that exceed these limits may be released by exception.

### 6.6.2 Specifications

The following recommended specifications are considered minimum requirements for the total microbial and total fungal counts.

Recommended Specifications

Materials of synthetic origin:

Total Aerobic Microbial Count: ≤ 500 cfu/g

Total Fungal Count: ≤ 50 cfu/g

Indicator Organisms: Based on finished product

Materials of natural origin:

Total Aerobic Microbial Count: ≤ 1,000 cfu/g

Total Fungal Count: $\leq$ 100 cfu/g

Indicator Organisms: Based on finished product

If an excipient or a drug substance has a compendial monograph or registration (approved NDA) requiring microbiological testing, those requirements will be met.

Packaging of excipients and drug substance must protect the materials from contamination during shipping and storage. Storage of excipients and drug substances must be under conditions which minimize bioburden proliferation. Materials which have microbiological specifications established are retested at specific intervals, such as semi-annually or annually, to confirm their microbiological quality and the acceptability of storage conditions.

### 6.6.3 Product Containers and Closures

Periodic monitoring of product contact containers and closures should be considered. The extent of monitoring, and specifications and/or control limits should be based upon the susceptibility of the product to microbial contamination and growth, such as aqueous based liquid products.

### 6.6.4 In-Process Materials

This is to establish maximum holding times prior to packaging. Unpreserved in-process materials that are stored for a period of time prior to use such as aqueous based liquid products, should be tested for their microbiological quality.

In addition, it may be necessary to monitor specific stages of the manufacturing process that have the potential to contribute bioburden to the finished product. During process validation, consideration should be given to evaluating the potential bioburden contribution from processes such as aqueous film coating solutions or suspensions which may be held over several days during use.

### 6.6.4.1 Water

Water is the most widely used solvent in the production, processing and formulation of pharmaceutical products. Several different grades of water for pharmaceutical applications exist. The quality attributes of water are mandated by the requirements of the application. Water is highly susceptible to microbiological contamination, and therefore, the system must be designed and validated with that in mind.

### 6.6.4.2 Drinking Water

Drinking water, or potable water used as source water for pharmaceutical process water, must comply with the quality attributes of the U.S. Environmental Protection Agency (EPA) National Primary Drinking Water Regulations (40 CFR 141) in which the level of coliforms are regulated. Several approved techniques to test for coliform bacteria are described in Standard Methods for the Examination of Water and Wastewater[3]. These techniques are the presence-absence test, most probable number (MPN) technique, chromogenic substrate coliform method and membrane filtration method.

Drinking water is source or feed water for pharmaceutical systems and typically contains a wide variety of microorganisms. The microorganisms may compromise subsequent purification steps and need to be controlled by system-design controls and sanitization to prevent biofilm formation and consequent planktonic, or free-floating, population. Heterotrophic plate counts should be performed on source water at regular intervals for information to monitor incoming bioburden and determine seasonal variation.

### 6.6.4.3 Purified Water, USP

Purified Water, USP is used in the production of non-sterile dosage forms, in equipment cleaning, and the preparation of some bulk pharmaceutical chemicals. It may be prepared by several methods such as filtration, deionization and reverse osmosis. Which ever method is chosen, the system must be validated and designed to limit microbial proliferation.

Recommended Action Limits and Alert Limits

Action Limits:
Total plate count: ≤ 100 cfu/mL
Absence of *Ps. aeruginosa* and *P. cepacia*

Alert Limit:
Total plate count: ≤ 50 cfu/mL

Trend analysis of microbiological data should be performed at regular intervals. Conclusions from these analyses should be used as a means of alert to signal the need for maintenance and/or sanitization. Alert and action levels should be established for monitoring and control of the water system. They should be established based upon system validation and monitoring data, system design specifications, product requirements and product quality concerns. The limits are established within process and product specification tolerances so that exceeding a limit does not imply that product quality has been compromised.

Objectionable organisms should be based upon the application of the water. For example, *Ps. aeruginosa* and *P. cepacia* are important organisms to consider for oral liquid manufacturing, but may not be a concern for oral solid dosage forms or bulk chemical processing. Although it may be desirable to have company-wide consistency in specifications, each site may have specific requirements based upon their application which need to be considered.

For Purified Water, USP, a minimum of 1.0 mL sample should be tested, preferably in duplicate or triplicate per sample site. Methods include pour plate and/or membrane filtration using cellulosic filters with no greater than a 0.45 μm porosity. At least once per year, isolates from system monitoring should be characterized to genus and species, where possible. Isolate characterization should be included in the system validation to determine baseline data. Methods used in system validation should not be changed unless thorough method validation and comparative studies are performed. Once justified, the method change should be documented via change control procedures.

As with other excipients, foreign compendial requirements need to be considered. However, for microbiological testing, the compendia provide little guidance and do not include microbiological specifications. When tested as described above, European requirements should be met.

High or low nutrient media may be used, but the incubation conditions must be tailored to the media. That is, if high nutrient media are used, such as plate count agar, the minimum incubation period should be 48 hours at 30 - 35 °C. If low nutrient media are used, such as R2A, lower incubation temperatures of 20 - 25 °C for longer times, e.g. 5 - 7 days, are necessary. Higher counts can be produced with low nutrient media, but the trade-off is longer incubation periods. The decision of which media to use should be made based upon the need for timely information and type of corrective actions when an alert or action level is exceeded.

Frequency of sanitization is generally determined by the results of system monitoring and documented in the water system SOP. The frequency of sanitization should be established such that the system operates in a state of microbiological control and does not routinely exceed alert limits. Frequency of monitoring should also ensure that the system consistently produces water of acceptable quality and is in control. Frequency is typically based upon system validation data and is documented in an SOP. Samples should be taken from representative locations from the pretreatment of system and distribution system. Pretreatment sites may be sampled less frequently than points of use. Pretreatment sites should be tested, at a minimum, on a quarterly basis. Source water should be tested, at minimum, on a quarterly basis. Each manufacturing point of use site should be tested, at minimum, on a monthly basis, with alternating sites being monitored in the distribution loop at least weekly.

Prior to sampling, points should be thoroughly flushed and sanitized. If hoses are typically used during routine applications, then sampling procedures should mimic those use procedures so that samples accurately represent use conditions. At locations where there is a potential for the presence of chlorine residuals, samples must be neutralized using agents such as sodium thiosulfate prior to microbiological analysis. Containers must be sterilized and aseptic techniques used. Sampling technique and container preparation must be documented in a Standard Operating Procedure (SOP).

## 6.7 Chiral Purity Using Chromatographic Techniques

In the mid 1980s, a new requirement was placed on the pharmaceutical industry for involving the development and testing of enantiomerically pure drug substances. As synthetic procedures are being developed and refined to make single-enantiomeric drug substances, analytical chemists have the challenge to develop methods for the separation and quantitation of low levels of undesirable enantiomers from the enantiomer having the desirable therapeutic activity. With the introduction of new chiral chromatography phases and chiral derivatizing reagents the technology for separating enantiomers has improved. At present there are a variety of techniques for developing methods for enantiomeric purity of drugs. The most common analytical techniques are:

- Use of a single chiral column to separate chiral and achiral related substances.
- Use of column switching techniques [5] to couple and uncouple chiral and achiral columns for efficient separation of the enantiomer, remove interfering components, pre-concentrate each component for sensitivity and obtain achiral and chiral results from single injection.
- Use of chiral mobile phase additives with an achiral column.
- Indirect separation of entiomers by LC. This is based on the simple chemical reaction of a mixture of enantiomers with an optically active pure chiral reagent as presented in the following simplified notation [6].

$$(R + S) + S' \rightarrow RS' + SS' \qquad (6.1)$$

where R and S are the enantiomers, S' is the chiral reagent and RS' and SS' are two diastereomers, which can be separated by a conventional LC system as two separate peaks. However, the above indirect mode of separation scheme becomes cumbersome and difficult if the chiral reagent S' contains possible chiral impurities R' in which case there will be four reaction products: RS', RR', SS' and SR', which will require a chiral LC column for separation.

## 6.8 Method Validation

Once an analytical method is developed for its intended use, it must be validated. The extent of validation evolves with the drug development phase. Usually, a limited validation is carried out to support an IND application and a more extensive validation for NDA and Marketing Authorization Application (MAA) submissions. Typical validation parameters recommended by FDA, USP and ICH are: accuracy, precision, linearity, specificity, limit of detection, limit of quantitation, ruggedness for drug substances and in addition to the above parameters, recovery and absence of matrix interference for dosage forms. Refer to Chapters 1, 2 and 3 of this book for further discussion on assay validation and Chapter 11 for validating cleaning methods. The discussion that follows will focus on validation requirements pertaining to characterization and quantitation of major components of bulk drug substances and finished products, methods for identifying and determining impurities and/or degradants in bulk substances and finished products.

Tables 1 and 2 list validation parameters which are typically evaluated for active drug substances and dosage forms for NDA and MAA submissions.

## 6.9 System Suitability

To ensure that any validated method will produce results that have the expected accuracy and precision on a given day, the FDA and USP have recommended the inclusion of system suitability tests as part of any LC system. As defined above, system suitability considers all the components of the hardware, i.e., sample and sample solvent column, mobile phase and equipment.

**Table 6.1**
Active Drug Substance Validation for NDA/MAA Submission

| | Validation Parameters Evaluated | | | | | | | | |
|---|---|---|---|---|---|---|---|---|---|
| Test | 1 | 2 | 3 | 4 | 5 | 6 | 7 | 8 | 9 |
| Identification (IR, NMR, MS) | | | | x | | | | | |
| Optical rotation | | x | | | | | | | |
| Particle size | | x | | | | | | | |
| Moisture by Karl Fischer | x | x | | | | | | | |
| Residual solvents (GC) | x | x | x | x | x | x | x | x | x |
| Impurities /degradants (LC or GC) | x | x | x | x | x | x | x | x | x |
| Assay (LC) | x | x | x | x | | | x | x | x |
| Assay (titration) | x | x | | | | | | | x |

Key
1 - Accuracy
2 - Precision (reproducibility/repeatability)
3 - Linearity
4 - Specificity/selectivity
5 - Limit of quantitation
6 - Limit of detection
7 - Ruggedness
8 - Recovery
9 - Solution stability

**Table 6.2**
Pharmaceutical Dosage Forms Validation for NDA/MAA Submission

| | Validation Parameters Evaluated | | | | | | | | |
|---|---|---|---|---|---|---|---|---|---|
| Test | 1 | 2 | 3 | 4 | 5 | 6 | 7 | 8 | 9 |
| Identification | | | | x | | | | | |
| Content uniformity | x | x | x | x | | | x | x | x |
| Dissolution | x | x | x | x | | | x | x | x |
| Related substances (LC) | x | x | x | x | x | x | x | x | x |
| Potency (LC/GC) | x | x | x | x | | | x | x | x |
| Preservative/ antioxidant | x | x | x | x | x | x | x | x | x |

Key
1 - Accuracy
2 - Precision
3 - Linearity
4 - Specificity
5 - Limit of quantitation[a]
6 - Limit of detection
7 - Ruggedness
8 - Recovery
9 - Solution stability

[a] A Practical limits of quantitation (PLOQ), which is set at 50% of the qualification limit (e.g. 0.05% if qualification limit is 0.10%) was also proposed by Wahlich at the 1995 November AAPS Annual Meeting in Miami, Florida.

System suitability criteria, which define the minimum acceptable performance prior to each analysis, can include resolution, precision, peak tailing factor, number of theoretical plates, capacity factor and relative retention. The most common measurements from the above list are system precision and system resolution between two closely eluting chromatographic peaks. The precision test concerns the repeatability of the system measurement analogous to the system reproducibility portion of the precision test. The latter is measured by injecting replicate aliquots of the same solution and peak heights, peak areas derived from the injections are determined and their RSD is the measure of the system's precision. The number of injections of the test solution required to determine the RSD depends upon the allowed precision level. It has been statistically determined [7] that, if the limit on the RSD is less than 2%, five replicate injections are required. If the limit is 2%-4%, six replicate injections are required. To minimize duplication effort analysts use the standard solutions prepared for the assay as the test solution for system suitability. In this way the same chromatograms may be used both for system-suitability calculations and for calculating the standard response factor.

The second part of the system suitability measurement is the system resolution, which is the calculation of the resolution between two closely eluting peaks, preferably peaks actually present in the test preparations. Thus the resolution of the system may be calculated from one of the same chromatograms run earlier to determine system precision. The resolution factor ($R_S$) is best calculated as per USP XXIII (see also Chapter 2):

$$R_S = \frac{2(t_2 - t_1)}{(w_2 + w_1)} \tag{6.2}$$

where $t_1$ and $t_2$ are the retention times of the two peaks and $w_1$ and $w_2$ are the widths of the extrapolated straight portions of the peaks and baseline.

The system suitability test should be performed each time a system is assembled for assay. If the system is dedicated for continuous use it will be sufficient to conduct an abbreviated check each day.

The Pharmaceutical Manufacturers Association (PMA) [8] carried out a survey on system suitability and instrument calibration.

There were 70 responders from industry (60% QC and 40% R&D). Highlights on system suitability survey follows:

- 100% use system suitability for assay and dissolution
- <50% use system suitability for impurities measurement
- Resolution, precision and peak symmetry are the most commonly measured and controlled parameters.

Seventy five percent responded that there are better ways for determining precision instead of five injections before use, such as interspersing standards throughout the LC run or bracketing standards.

All responders agreed that system suitability should be tested before each run or before each change to the system. A run is considered as a set of samples and standards prepared and analyzed in a continuous fashion with the system suitability test being conducted before each run. This applies both to dedicated equipment as well as to equipment set up for a specific analytical run.

Wahlich [9] has proposed system suitability tests (SSTs) to be applied as a check during the validation of a chromatographic method. Table 3 lists each of the validation parameters which are measured during the validation of the method, listing the number of gaps in what is presently being done. Suggestions are made of how these gaps might be filled. In addition, the usefulness of SSTs for peak tailing and column efficiency is questioned.

## 6.10 Analytical Support for Safety Assessment Studies

Analytical support for non-GLP and GLP safety-assessment studies is required during the course of drug development. These methods are necessary to determine the concentration, stability and homogeneity of the test article in the selected vehicle. For the most part the vehicles used are saline, methyl cellulose, tween, propylene glycol and dimethylsulfoxide. For long term studies dry feed is used because of stability considerations.

**Table 6.3**
Method validation parameters and system suitability tests applied

| Validation parameter | Traditional SSTs | Recommended SSTs |
|---|---|---|
| Ruggedness/ robustness | None | Check on critical method parameters |
| Accuracy | None | Control samples<br>Re-extraction<br>Mass balance |
| Precision | RSD of replicate injections | RSD of replicate injections<br>RSD of replicate sample preparations |
| Selectivity | Resolution check | Resolution check (using impure standards or samples of the impurities) |
| Stability of measurement solutions/ systems | None | Comparison of standards at start and end of run |
| Linearity | None | Use of standards at different concentrations |
| LOD/ quantitation (signal-to-noise ratio) | None | Calculation of signal to noise ratio |
| General acceptability | None | Chromatogram compared to reference chromatogram |
| None | Tailing factor/peak asymmetry | None |
| None | Column efficiency/plate count | None |

Typically LC is used for determining formulation stability, test article concentration and for establishing homogeneity of suspensions. Methods validation is usually performed over a range of 50-150% using an acceptance criteria for recovery not less than 98% for solutions and suspensions and not less than 90% for feed. The most common assay methods are isocratic LC methods. Formulation specifications for label strength varies from 90-110% to 80-120% depending on the type of feed. Many firms do not have specifications for degradation products but will evaluate each formulation on a case-by-case basis.

All equipment used to support GLP studies undergo full calibration at regular intervals. Prior to initiation of a non-clinical laboratory study, Analytical R&D develops methods to:

- Determine identity, strength, purity and composition for the test or control article
- Evaluate the stability of the test or control article under use conditions
- Determine the uniformity of the control article

## 6.11 Excipients

Excipients are substances, other than the active drug, which have been appropriately evaluated for safety and are included in a drug delivery system to either:

- aid in processing the formulation during its manufacture, or
- protect, support or enhance stability, bioavailability or patient acceptability

Analytical testing of excipients usually follows the applicable compendia. Scientifically sound alternate test methods are also acceptable. However, in the event of a dispute the compendial method is applied as the reference test by the FDA. It is also important to set appropriate limits for impurities based on appropriate toxicological data or compendial requirements. Since many excipients are purified using organic solvents, it is important to include limits for the residual solvents as part of excipient specifications.

## 6.12 Reference Standards in Pharmaceutical Analysis

There is no consensus amongst pharmaceutical companies [10] as to the phase in drug development process which requires the establishment of a reference standard as defined by the FDA [11]. However, most companies designate a drug substance lot as a standard soon after a drug candidate has been identified or when GLP investigations have been initiated. Some companies use representative lots of ADS for the first standard. The chromatographic purity of the standard is calculated using the relationship [100% - total related substances by LC] as the basis for the purity value. The individual related substances in the standard are quantitated using gradient elution versus ADS standards, if available, prepared at 0.1-1.0% of the sample concentration. Since the related substances may have different molar absorptivities than the ADS at the chosen wavelength, an effort should be made to select an appropriate wavelength using a diode array detector to minimize these differences. To assign "purity" or "use as value", corrections must be made for the amount of residual organic solvents and water present in the standard.

Concurrent with drug development phase, the quantity and preparation of standards is also improved. By phase 2-phase 3, reference standards are prepared by successive recrystallization of a released lot of bulk drug substance until a chromatographic purity of much greater than 99.0% is achieved.

The standard is thoroughly characterized, by elemental analysis, ion chromatography, nuclear magnetic resonance, etc., in addition to the tests and methods used to control the drug substance. Absolute purity measurements such as phase solubility or differential scanning calorimetry is used to assign a purity value. Some companies assign a purity value obtained by subtracting the related substances, water and the residual solvents from 100% as discussed above. Such standards are labeled "primary reference" or "analytical reference standards". These standards are used for structure elucidation as part of an NDA. It is recommended that such standards be stored under a controlled environment (i.e. sealed ampules) in very small quantities. The analytical reference material described above is maintained as the standard (benchmark) against which "working standards" are judged.

A working standard is a bulk drug substance which has been tested and released by the tests and methods used to control the drug substance. This material is analyzed against the "primary standard" and a purity value is assigned. The use of a working standard is common but is not practiced by all pharmaceutical companies.

Retesting schedules for reference standards vary from company to company. Yearly retesting for new compounds and longer retesting intervals with stability justification is the usual practice. Generally, only degradants and other attributes which are expected to change with time such as moisture and residual solvents are performed for retest.

Finally, it is not always necessary for the reference standard to be in the same form as the corresponding bulk drug. Often, for improved stability the free base of the drug substance is used.

Authentic standards for related substances, i.e., degradants and synthetic impurities, are usually certified by measuring the chromatographic purity using gradient elution, by thermal gravmetric analysis for quantitating the residual solvents and by differential scanning colorimetry for thermally stable materials. To conserve the supply of such standards, response factors are developed as per 6.3.1.4c of this chapter.

## 6.13 Conclusions

Drug development is a dynamic process. As such, testing requirements and controls should be designed to parallel the development phase to support ongoing clinical trials. With the ever increasing costs of drug development, it is important for the analysts to keep this in mind and avoid unnecessary and cumbersome measurements early on in the development process. Good science, careful planning, good documentation of all data and procedures should satisfy most regulatory agencies. In the final analysis, this will reduce regulatory review time and expedite approval of a product.

## Acknowledgements

The author gratefully acknowledges Kathy Sylvester for her assistance in preparing this chapter and Dr. Christopher M. Riley and Dr. Thomas W. Rosanske for their critical review and comments.

## References

1. International Conference on Harmonization: Guideline on Validation of Analytical Procedures: Definitions and Terminology Avaliability, *Fed. Reg.,* **60**, 11260-2 (1995)

2. M.P. Newton, J. Mascho, R.J. Maddox, *ACS Symposium Series* **512**, 41-53 (1992)

3. Jeri May, DuPont Merck Pharmaceutical Company, personal communications (1995)

4. A.E. Greenberg, L.S. Clesori and A.D. Eaton. Standard Methods for the Examination of Water and Wastewater, 18th ed. American Public Health Association, Washington, DC (1992)

5. R. Bopp, "Analytical Technology Relevant to Stereoisomerism", AAPS Workshop on Impurities in Drug Substance and Products, April 3-4, Arlington, VA (1995)

6. W. Linder, C. Petterson in "Liquid Chromatography in Pharmaceutical Development: An Introduction", I.W. Wainer (ed.), Aster, Springfield, OR, 63-131 (1985)

7. E. Debesis, J. Boehlert, T. Givand and J. Sheridan *Pharm. Tech* **Sept**., p. 120-137 (1982)

8. PMA (Analytical R&D Subsection and Pharmaceutical R&D subsection) meeting Washington, DC, April 10-12, 1994

9. J.C. Wahlich and G.P. Carr, *J. Pharm. Biomed. Anal.,* **8**, , 619-623 (1990)

10. PMA Workshop on Reference Standards, sponsored by the PMA Analytical Steering Committee, Washington DC, 1992

11. FDA Guidelines for Supporting Documentation in Drug Applications for the Manufacture of Drug Substances, February 1987

## Chapter 7

# Dissolution Studies

*Thomas W. Rosanske and Cynthia K. Brown*

## 7.1 Introduction

Dissolution testing monitors the rate at which a solid or semi-solid pharmaceutical dosage form releases the active ingredient(s) into the liquid medium under standardized conditions of liquid/solid interface, temperature, and media composition. Scientists have been conducting dissolution studies for many years. However, it was not until 1970 that dissolution testing was officially recognized as a product quality indicator when it was incorporated into twelve monographs in the United States Pharmacopeia/National Formulary (USP/NF), USP XVIII/NF XIII [1]. In the current U.S. Pharmacopeial edition, USP 23/NF 18, nearly all solid dosage form monographs include a dissolution test [2]. Dissolution tests are defined as Category III by the USP, i.e., "Analytical method for the determination of performance characteristics..."[3]. In spite of the now general acceptance of dissolution testing of pharmaceutical products as a standard practice, the value of the dissolution test as applied to pharmaceuticals remains a topic of much discussion and controversy among pharmaceutical scientists. Investigators have for many years attempted to justify dissolution testing on *in vitro/in vivo* correlations but have met with only moderate success. However, when dissolution is the rate limiting factor in the absorption of the drug, a rank order correlation between dissolution rate and bioavailability often can confirm bioequivalency among product batches [4]. Dissolution testing is primarily used in industry as a quality control tool to monitor the formulation and manufacturing process of the dosage form. Dissolution is considered by most regulatory agencies as a highly critical quality characteristic for most solid dosage forms.

It is not the primary intent of this chapter to discuss in detail the theory and development of dissolution tests, but rather the validation exercise. For a complete and excellent discussion on the theory of dissolution as it relates to dosage forms and bioavailability characteristics, the reader is referred to a book compiled by Abdou [5]. In order to validate a dissolution method, however, it is important to have a good understanding of the theory of dissolution and the roles of the key parameters of a dissolution test. A complete dissolution validation package would consider at a minimum the dissolution apparatus used, calibration requirements and any appropriate governmental or regulatory guidelines. Therefore, it is important to address these issues here in the context of dissolution test validation.

## 7.2 The Dissolution Test

In order for a dissolution test to demonstrate the unique dissolution characteristics of the dosage form, the dissolution procedure should be based on the physical and chemical properties of the drug substance as well as the dosage form characteristics. Some of the physical-chemical properties of the drug substance which influence the dissolution characteristics are:

- solubility in water and other solvents
- effect of pH on solubility-ionization constants
- solution stability
- particle size/surface area
- crystal form
- common ion effects
- ionic strength
- buffer effects
- octanol/water partition coefficients
- effect of temperature on solubility

Once the drug substance properties have been determined, the actual dosage form needs to be considered. For example, is the dosage form a tablet, capsule, semi-solid (ointment or cream), or transdermal and is the dosage form designed for immediate release or controlled release of the drug product? Of key importance is the potency of the dosage form or the amount of drug to be delivered and the rate at which the drug is to be delivered. This relates directly to the mathematical expression of dissolution rate which is defined by the Noyes-Whitney equation:

$$\frac{dW}{dt} = k_1 S(C_{sat} - C_{sol}) \qquad (7.1)$$

where:

$\frac{dW}{dt}$ = the dissolution rate

$k_1$ = a dissolution constant

$C_{sat}$ = the concentration of a saturated solution

$C_{sol}$ = the concentration of the solution at any given time

S = the surface area of the solid

*In vivo*, the gastrointestinal tract acts as a natural sink, i.e., the drug is absorbed as it dissolves. *In vitro*, sink conditions are simulated by using either a large volume of dissolution medium or by replenishing the medium with fresh solvent at a specific rate. (By keeping the volume of dissolution medium at least five to ten times greater than the saturation volume, sink conditions are approximated [6])[1]. When sink conditions are achieved, $C_{sat} >> C_{sol}$, the equation simplifies to:

$$\frac{dW}{dt} k_2 S \qquad (7.2)$$

where:

[1] According to the USP <1088>, the quantity of medium used should be not less than three times that required to form a saturated solution.

$\frac{dW}{dt}$ = the dissolution rate

$k_2$ = a dissolution constant

S = the surface area of the solid

In this case, the dissolution rate is characteristic of the release of active ingredient from the dosage form rather than the solubility in the dissolution medium. Therefore, sink conditions are one of the main experimental parameters to be controlled in dissolution testing. Sink conditions can be achieved by the appropriate selection of the dissolution apparatus and dissolution medium. Based on drug substance and dosage form characteristics, the appropriate dissolution apparatus should be selected.

### 7.2.1 Apparatus

For a dissolution test to be used universally to control the consistency of a pharmaceutical dosage form, some controls must be placed on the type of apparatus used. The USP XVIII and NF XIV described only three types of apparatus [1], but more recent developments in dissolution testing have resulted in seven types of apparatus approved for use in the USP 23/NF 18 [7]. USP apparatus Type 1 and Type 2 are by far the most frequently used for immediate release and most sustained release dosage forms. These types are described in some detail below. USP Type 4 in used rather infrequently in the United States, but finds more use in the European community. USP Types 3, 5, 6, and 7 are much less frequently used and only a brief description is given below. Compendia other than the USP, e.g., the Pharmacopoeia of Japan (JP) [8], the British Pharmacopoeia (BP) [9], and the European Pharmacopoeia (EP) [10] all contain the same basic Type 1 and Type 2 dissolution equipment as described in the USP. The BP and EP also include a flow-through apparatus identical to the USP Type 4 apparatus.

### a) USP Type 1

This apparatus, shown schematically in Figure 1, consists of a covered vessel of specified shape and dimensions and capacity of 1000 ml, a metallic shaft one end of which attaches to a motor, and a cylindrical metallic mesh basket that attaches to the opposite end of the shaft. The dosage form is placed inside of the basket and the basket assembly is immersed in the dissolution vessel containing a nominal volume of dissolution medium and rotated at a specified speed.

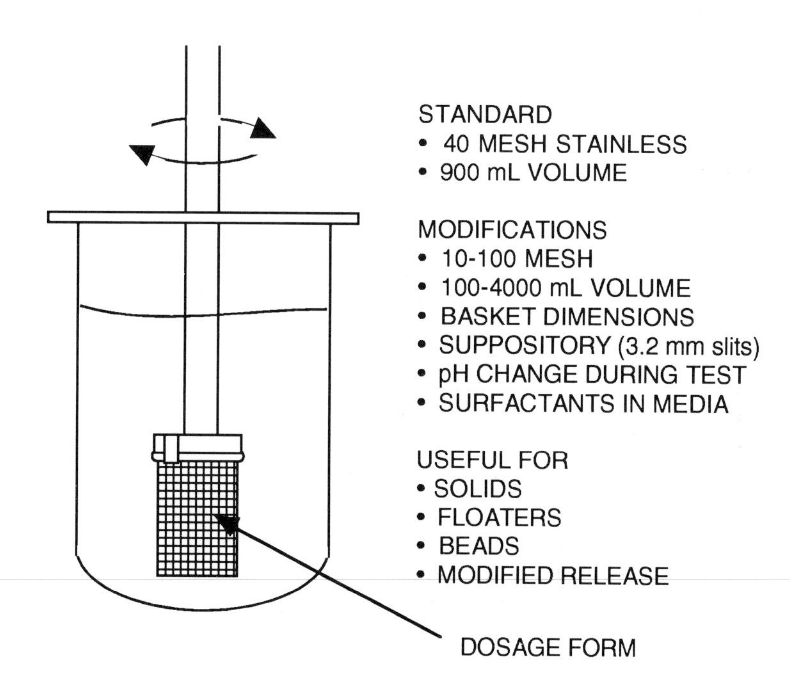


**Figure 7.1**
USP Apparatus 1, Rotating Basket. [Reproduced with permission of Royal Hanson]

### b) USP Type 2

The USP Type 2 is currently the most frequently used apparatus for all solid dosage forms. The dissolution vessel used with this apparatus is the same as for the USP Type 1 apparatus. However, the basket assembly is replaced by a paddle of specified dimensions, shown schematically in Figure 2. With this apparatus, the dosage form is dropped directly into the vessel containing the dissolution medium and allowed to sink to the bottom; the paddle is then rotated at a specified speed but can be immersed in the vessel when the dosage form is dropped as long as the rotation device is switched off, which is the common industry practice. The USP specifies to place the dosage form in the apparatus and immediately operate at a specified speed. The USP does not state whether the paddles can or cannot be immersed prior to addition of the dosage form, only that the rotation be started after the dosage form has been added.

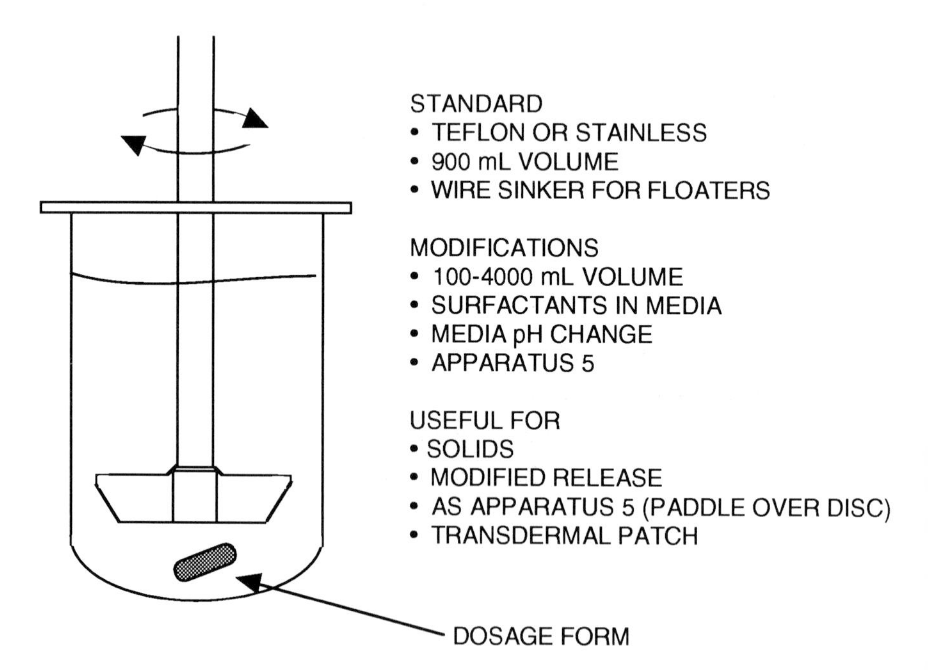


**Figure 7.2**
USP Apparatus 2, Paddle. [Reproduced with permission of Royal Hanson]

## c) USP Type 3

The USP Type 3 apparatus finds its primary use in extended release formulations, although it is not used extensively. It consists of a cylinder with a screen on one end that contains the dosage form and a larger glass cylinder into which the dissolution medium is placed. The dosage form-containing cylinder is immersed in the dissolution medium and agitated with vertical strokes.

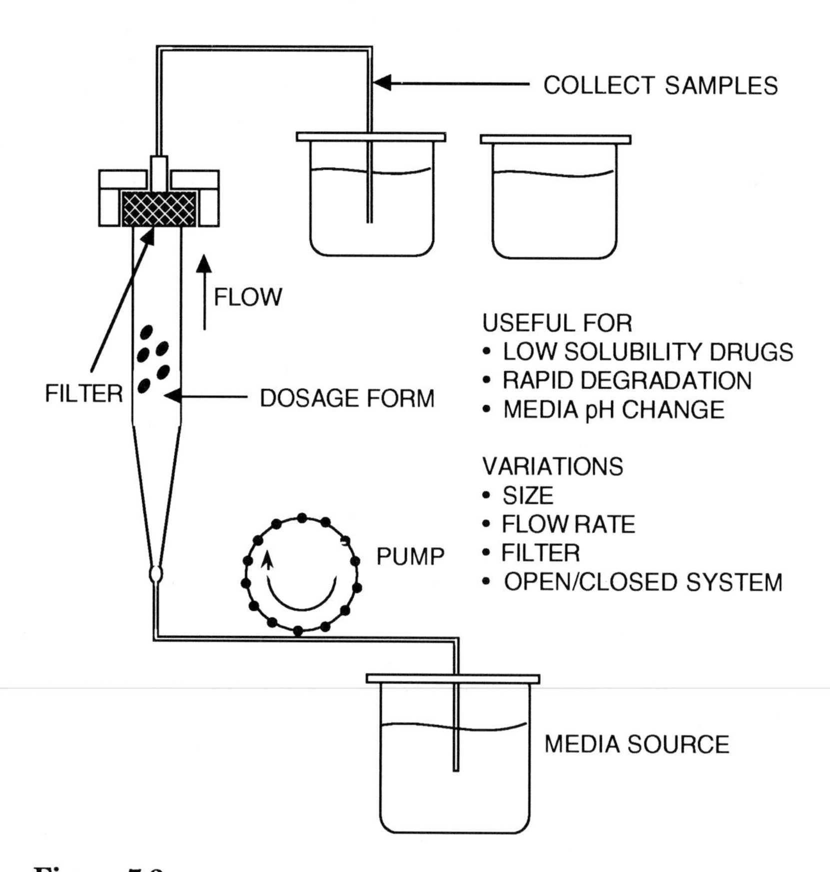


**Figure 7.3**
USP Apparatus 4, Flow-through Cell. [Reproduced by permission of Royal Hanson]

#### d) USP Type 4

Figure 7.3 shows a schematic diagram of USP Apparatus 4. This apparatus consists of a flow through cell into which the dosage form is placed. Dissolution medium is continually pumped through the flow cell and the eluent analyzed for drug content. This apparatus finds its greatest utility in the testing of poorly soluble drugs as sink conditions can be maintained by continually passing fresh dissolution fluid over the dosage unit. This apparatus finds more widespread use in Europe than elsewhere.

#### e) USP Types 5, 6 and 7

The USP Types 5, 6, and 7 systems are all used in the testing of transdermal dosage forms. USP Type 5 is a modification of the Type 2 system and uses a rotating paddle over a disk assembly. The USP Type 6 uses the same assembly as Type 1, with the transdermal dosage form attached to the bottom of the shaft. The USP Type 7 apparatus consists of a reciprocating disk assembly whereby the dosage form is placed on a cylinder and vertically reciprocated in dissolution medium.

### 7.2.2 Sampling and Analytical Instrumentation

As a single dissolution test generally requires the sampling and analysis of several samples (usually six per sampling interval), it is necessary to have efficient sampling systems and rapid analytical systems in place. Sampling can be performed manually or by using automated sampling systems. Manual sampling can be quite labor intensive and tedious. The manual sampling from six vessels for a single time point in a dissolution test can take several minutes making the establishment of dissolution profiles for rapidly dissolving dosage forms quite difficult at early time points. Several automated sampling systems are commercially available in today's market. The automated sampling systems are typically microprocessor or computer controlled allowing for precisely timed sampling at frequent intervals, if necessary. Some automated sampling systems transfer the samples to collection tubes for manual transfer to the analytical instrument while other sampling systems transfer the samples directly into the analytical instrument for analysis. It is important to note that all sampling systems should be evaluated for absorption to tubing or filters. Carry-over between samples should also be determined, especially when common sampling pathways are used. Errors associated with automated sampling systems are usually related to partial or complete blockage of the sampling lines. Therefore, it is recommended that the appropriate flow rate be determined and used prior to each use of an automated sampling system.

Because of the large number of samples normally generated in a dissolution test, the analytical system should be relatively rapid, allowing for a high throughput of samples. The most common analytical instruments used for dissolution testing are UV-Visible spectroscopy and liquid chromatography with UV detection (LC-UV).

UV-Visible spectroscopy allows for rapid analysis of samples. Diode array based instruments with sophisticated computer generated data analysis have improved the quality and speed of single component as well as multicomponent analysis by UV-Visible spectroscopy. When this mode of analysis is used, the accuracy should be confirmed by a selective mode of analysis such as LC. LC analysis is a more discriminating and selective means of analysis. When LC analysis is required, a rapid chromatographic method is desirable for the assay. It is important to note that the analytical instrumentation should be checked for wavelength accuracy and repeatability as well as photometric accuracy and repeatability.

### 7.2.3 Single-Point Test *versus* Dissolution Profile

The sampling and analysis procedures used for a dissolution test may to a large degree depend on the type of dosage form and purpose of the test. Ideally, the *in vitro* dissolution rate for a given formulation and dosage form will be reflective in some way of *in vivo* availability of the drug, thus allowing for establishment of a correlation between the *in vitro* dissolution behavior and one or more pharmacokinetic parameters ($C_{max}$, $t_{max}$, AUC, etc.). A well established correlation will allow for a reasonable prediction of the *in vivo* behavior of formulations without performing a bioavailability study. Finding the appropriate correlation, if one exists, has been the focus of numerous studies [11,12]. For many products, good correlations have been obtained, but several other products exist for which no suitable correlations have been found [12]. The establishment of *in vitro/in vivo* correlations has been the subject of much discussion in recent years. The USP has defined 3 levels of correlation approaches with varying degrees of usefulness [13].

It is generally more difficult to generate useful or predictive correlations between measured *in vitro* dissolution rate and bioavailability for immediate release dosage forms. Since dissolution rates for such products are by design relatively rapid, it is often found that dissolution of the drug may not be the rate limiting factor for *in vivo* activity. For immediate release dosage forms, only single point dissolution information is usually generated. Investigators have generally found that for such dosage forms a minimum dissolution rate (single point) could be found beyond which all products would show equivalent *in vivo* behavior [14].

In contrast to immediate release formulations, extended or controlled release formulations are designed to release drug from the product matrix over an extended period of time, generally 12 to 24 hours or longer. This implies that the rate at which a drug dissolves from the formulation matrix is a controlling factor in the bioavailability of the drug. For such products, it is critical to establish a correlation at multiple time points in order to ensure batch to batch product consistency over the entire time course of the product. Thus, it is obvious that single point dissolution tests are not adequate for product control, and dissolution profiles are necessary.

The most useful approaches to *in vitro/in vivo* correlations for extended release products are to use either statistical moment analysis to obtain and compare *in vitro* and *in vivo* mean residence times or mathematical deconvolution techniques to generate and compare *in vitro* and *in vivo* dissolution profiles [13]. The latter of these is preferred in most situations as it represents a point to point correlation. Correlations established with this technique should result in the *in vitro* data being highly predictive of *in vivo* performance.

## 7.3 Calibration

In order for a dissolution method to be considered valid, the dissolution apparatus must be set-up, calibrated, and operated in compliance with the appropriate compendia, if applicable. USP 23/NF18 general chapters on *Dissolution* <711> and *Drug Release* <724> list the apparatus specifications, the apparatus suitability test (calibration requirements), the dissolution medium requirements, as well as specific procedure requirements for USP apparatus Types 1-7 [7].

The system suitability of Apparatus 1, 2, and 4 as described in USP 23/NF18 requires the testing of two types of USP calibrator tablets, non-disintegrating salicylic acid tablets and disintegrating prednisone tablets. The system suitability of USP Apparatus 3, the reciprocating cylinder, requires the testing of chlorpheniramine maleate extended-release tablets and theophylline extended release beads [15]. The calibrator tablets are supplied with a certificate of acceptable ranges established for the specific lot of calibrator tablet. The appropriate USP reference standard should be used when testing the calibrator tablets. Calibration should be conducted at regular intervals (usually every six months) or when the apparatus is moved or maintenance modifications are performed. For Apparatus 1 and 2, the acceptance ranges for the calibrator tablets are given at both 50 and 100 rpm. Some dissolution tests require modifications in the set-up of the Apparatus 1 and 2 such as a two liter vessels or rotation speeds other 50 or 100 rpm. In these situations, if the apparatus passes the established calibration tests, it is considered suitable for the atypical conditions. Apparatus 5 (paddle over disk) is also modification of Apparatus 2, therefore it does not require any additional calibration [15].

Currently, the use of the USP calibrator tablets to certify dissolution apparatus suitability is a controversial topic in the pharmaceutical industry. While the rationale behind the use of the USP dissolution calibrator tablets and the interpretation of calibrator tablet failures is beyond the scope of this chapter, the reader is referred to discussions on these topics in the Pharmacopeial Forum [15, 16, 17].

## 7.4 Regulatory Guidelines

The regulatory agencies for the various global regions generally address dissolution guidelines in terms of the particular testing necessary to demonstrate the appropriate or intended release from a dosage form. These guidelines relate more to the development of an appropriate dissolution method than the actual validation procedure. The specifics of the analytical validation for dissolution procedures are not as a rule separated from discussions of general method validation as most of the critical validation analysis parameters do not differ between dissolution methods and, for example, assay methods. As the regional regulatory and ICH guidelines are discussed in detail in Chapter 3, the reader is referred to that chapter.

## 7.5 Analytical Validation

Once the appropriate dissolution conditions have been established, the analytical method should be validated for linearity, accuracy, precision, specificity and ruggedness. Each of these analytical parameters have been discussed in detail in previous chapters. This section will discuss these parameters only in relation to issues unique to dissolution testing. All dissolution testing must be performed on a calibrated dissolution apparatus meeting the specified mechanical standards.

### 7.5.1 Linearity

Detector linearity should be checked over the entire range of concentrations expected during the procedure. For instant release formulations, a concentration range of at least 50% of the lowest concentration expected in the dissolution vessel to 120% of the highest concentration is sufficient. For controlled release products, the concentration range should extend from approximately 10% to 120% of that expected from dosage strength. If a controlled release product is formulated in multiple strengths, the detector linearity should be confirmed from 10% of the lowest concentration to 120% of the highest concentration. Typically the concentration range is divided into five evenly spaced concentrations. Linearity testing of the dosage form should cover the entire specification range of the product. All samples should be heated to 37°C (or the specified dissolution temperature); this is especially important if the samples are close to sink conditions.

### 7.5.2 Accuracy

Samples are prepared by spiking bulk drug and excipients in the specified volume of dissolution fluid. The concentration ranges of the bulk drug spikes are the same as those specified for linearity testing. If the dosage form is a capsule, the same size and color of capsule shell should be added to the mixture. The solutions should be tested according to the parameters specified in the method, i.e., temperature, rotation speed, filters, sampling mode, and detection mode. A mixture of dissolution fluid and the excipients (including the capsule shell, if applicable) should be tested to determine any excipient interference. Accuracy should be determined at each specification interval for the dosage form. Detector linearity can be determined from this same set of samples.

### 7.5.3. Precision

Precision is determined by testing at least six aliquots of a homogenous sample of each dosage strength according to the dissolution method. The precision should be assessed at each specification interval for the dosage form. If the dosage form requires the use of a sinker, the sinker specified in the method should be used in precision testing.

### 7.5.4 Specificity

The dissolution analysis method must be specific for the bulk drug substance in the presence of placebo. The accuracy testing should establish specificity. To establish appropriate specificity, the accuracy sample solutions should be monitored for degradation. Simply monitoring the UV spectra of the solutions is not sufficient in determining degradation since many degradation products will have the same UV spectrum as the parent compound. Therefore, specificity testing should be confirmed by analyzing accuracy samples with a selective analysis mode such as LC. If the capsule shell interferes with the bulk drug detection, the USP allows for a correction for the capsule shell interference. Corrections greater than 25% of labeled content are unacceptable [18].

### 7.5.5 Robustness/Ruggedness

Robustness testing should determine the critical parameters for a particular dissolution method. By subjecting each dissolution parameter to slight variations, the critical dissolution parameters for the dosage form will be determined. This will facilitate method transfer and troubleshooting. Robustness testing should evaluate the effect of varying media pH, media volume or flow rate, rotation speed, sample position in the apparatus, sinkers (if applicable), media deaeration, temperature, and filters. Ruggedness of the methods should be evaluated by running the method with multiple analysts on multiple systems. If the analysis is performed by LC, the effect of columns and mobile conditions should also be addressed.

### Acknowlegement

The authors wish to acknowledge Diane L. Peterson (Hoechst Marion Roussel) for providing much useful information in the form of references and consultation, and Timothy McCormick (DuPont Merck) for his constructive review of this chapter and valuable suggestions for revision.

### References

1. United States Pharmacopeia and National Formulary, United States Pharmacopeial Convention, Rockville (1970)

2. United States Pharmacopeia 23/National Formulary 18, United States Pharmacopeial Convention, Rockville (1995)

3. United States Pharmacopeia 23/National Formulary 18, United States Pharmacopeial Convention, Rockville, pp. 1983-1984 (1995)

4. W.A. Hanson, *Handbook of Dissolution Testing*, 2nd edition, revised, Eugene, OR, Aster, p. 14 (1991)

5. H.M. Abdou, *Dissolution, Bioavailability, & Bioequivalence*, Easton, PA, Mack (1989)

6. W.A. Hanson, *Handbook of Dissolution Testing*, 2nd edition, revised, Eugene, OR, Aster, pp. 21-22 (1989)

7. United States Pharmacopeia 23/National Formulary 18, United States Pharmacopeial Convention, Rockville, pp. 1791-1799 (1995)

8. The Pharmacopoeia of Japan, twelfth edition, The Society of Japanese Pharmacopoeia, pp. 27-28 (1992)

9. British Pharmacopoeia, Her Majesty's Stationery Office, London, pp. A160-A162 (1993)

10. European Pharmacopoeia, Maisonneuve, Sainte-Ruffine, pp. v.5.4.-1-v.5.4.-8 (1992)

11. V. Banakar and L. Block, *Pharm Tech*, **7**, 107 (1983)

12. H.M. Abdou, *Dissolution, Bioavailability, & Bioequivalence*, Easton, PA, Mack, pp. 502-512 (1989)

13. United States Pharmacopeia 23/National Formulary 18, United States Pharmacopeial Convention, Rockville, pp. 1928-1929 (1995)

14. L.J. Leeson, 29th Annual Arden House Conference, Harriman, NY (1994)

15. L.T. Grady, *Pharm. Forum*, **20:6**, 8567-8569 (1994)

16. V.A. Gray, B.B. Hubert, and J.A. Kraskowski, *Pharm. Forum*, **20:6**, 8571-8573 (1994)

17. S.A. Qureshi and I.J. McGilveray, *Pharm. Forum*, **20:6**, 8565-8566 (1994)

18. United States Pharmacopeia 23/National Formulary 18, United States Pharmacopeial Convention, Rockville, pp. 1793-1795 (1995)

# Chapter 8

# Robotics and Automated Workstations

*Julie J. Tomlinson*

## 8.1 Introduction

Laboratory automation is widely used in the pharmaceutical industry to support quality control, nonclinical and clinical drug development, and drug discovery. Analytical applications for which automation is commonly utilized include dissolution tests, content uniformity and tablet assays, stability tests, aerosol dose tests, biological fluids analyses, toxicological feeds analyses, immunological assays, compound syntheses, and receptor-binding assays.

Pharmaceutical laboratory applications that utilize laboratory automation can have widely differing means of verifying quality of analytical performance. Laboratories that fall under the jurisdiction of GLP or cGMP regulations commonly use instrument calibrations and validations to verify and monitor analytical performance. Many of the principles behind calibration and validation of automated processes are the same as those for manual processes. However, some aspects of the implementation of calibration and validation procedures are unique to automation.

### 8.1.1 Scope

Laboratory automation is used for preparation, introduction, and analysis of samples, as well as collection, analysis, and reporting of results. This chapter specifically focuses on validation of automated workstations and robotic systems that are used for sample preparation and introduction. The validation practices described in this chapter pertain to laboratories that follow guidelines described in Good Laboratory Practices (GLPs). Procedures in laboratories that follow current Good Manufacturing Practices (cGMPs) are more briefly described.

### 8.1.2 Definitions

Pharmaceutical laboratory automation most commonly exists as flexible automation. This is particularly true in research and development where the compounds analyzed and the techniques and methods used to analyze them frequently change. Automation of these applications requires flexible hardware and software to be useful and cost-effective in the changeable work environment.

A *robot* is a mechanical arm connected to a microprocessor-driven controller that is used to program its movements and positions. A robot arm may have a hand or gripper attached to it, and it typically moves within a cylindrical, revolute, or Cartesian coordinate system envelope. A robot may have a variety of different sensors or sensing capabilities that provide intelligence about its positions and its environment. The arm is surrounded by workstations, supplies, and other equipment that it requires to perform its work. Collectively, the robot and its accompanying equipment are called a *robotic system*. Robotic systems are typically used to perform complex tasks that have multiple pick-and-place manipulations. Specific details about robotic arms, controllers, sensors, and system hardware and software may be found in a variety of literature [1-3].

An *automated workstation* is very similar in function to a robotic system, and it is often used to automate processes that require fewer manipulations than robotic system processes. It is typically faster, smaller, and easier to install and operate than a robotic system. A workstation requires less sensory capability because its environment and positions are less likely to encounter frequent variation. Certain standard components can be added or removed, and the software is often menu-driven rather than fully programmable. As with any piece of equipment, a workstation can be customized to meet analytical requirements and specifications, which can significantly increase its complexity.

Because the sample preparation operations they perform are very similar, the calibration and validation requirements of robotic systems and workstations are also very similar. Therefore, robotic systems and automated workstations are treated equally in this text unless otherwise specified.

### 8.1.3 Planning an Automated Laboratory

Even a cursory review of current literature shows that all of the major pharmaceutical companies in the United States use automation to achieve their laboratory goals. Though their priorities differ, most have chosen to invest in automation technology for one or more of the following reasons:

- to increase analytical productivity
- to improve analytical precision and accuracy
- to retain personnel by minimizing analytical drudgery
- to separate personnel from work place health and safety hazards
- to help document sample history

Automation does not decrease the need for personnel, though once it is reliably implemented in a laboratory, the amount of work that can be performed by a person typically increases per unit of time [4, 5].

When first planning to implement automation, it is often desirable to automate the most complex laboratory tasks. However, because automation frequently adds a dimension of complexity that can prove overwhelming, this approach can lead to failure when complex methods are implemented. Even seemingly simple automated workstations can be fraught with unexpected implementation problems when users try to validate their assays. It is prudent, therefore, to automate and validate simple routine processes until the expertise that is required to successfully program and engineer complex systems is developed. It is equally important to develop methods and processes that are specifically automation-compatible.

Automation often changes the types of skills that are needed to successfully develop and implement a reliable analytical method [5, 6]. The types of skills that are required for success include: 1) operators who can routinely load samples, reagents, and disposables onto the system, start automated runs, and document and report analytical results; 2) robotic application developers who typically develop or co-develop the automated chemistry and set up the application; and 3) engineers who are available to repair, customize, or develop hardware components and interfaces. For complex robotic applications, one operator usually supports one robotic system. One applications developer and one engineering group often support many automated systems.

Suitable laboratory space is also a prerequisite to successful system implementation. The design of a fully-automated laboratory can be quite different from a traditional laboratory and many interesting laboratory designs have arisen from the need to accommodate automation [5, 8-10]. These laboratories have specific key features, such as: a) vast, open space to accommodate large systems and associated equipment; b) well-protected, isolated electrical circuitry; c) centralized, continuous gas supplies; d) customized ventilation systems; and/or d) numerous utility connections in floors, from ceilings, or from posts. These and many other features are often needed to smoothly install automated systems. It is prudent to consult literature, scientists who currently use automated systems, and system manufacturers to plan the laboratory well before installing automated equipment. Failure to do so can result in very costly and time-consuming laboratory retro-fitting and analytical down time.

A frequently overlooked requirement for successful automation lies in managing the behavioral and cultural aspects of implementation. The entire laboratory environment may change, such as scientists' daily activities, their tools, and their workspaces. Requirements for success in implementing laboratory automation closely parallel those described for organizational reengineering processes. Behavioral and cultural requirements include senior management leadership, commitment to supplying necessary resources, thorough training, setting realistic goals, and so on [7]. Sincere attention to these issues is required for successful automation.

## 8.2 Validation Requirements: Chemistry or Computer?

Many laboratory automation scientists have grappled with the issue of whether computer validation regulations are applicable to laboratory robotics and automated workstations. The United States

Environmental Protection Agency (US EPA) issued a Good Automated Laboratory Practices (GALP) document in 1990 that applies to all laboratories that provide data to US EPA [11]. This document gives thorough recommendations for implementing and operating an automated laboratory; however, each section of the GALP manual specifically states that its recommendations apply to automated data collection systems and Laboratory Information Management Systems (LIMS). No mention is made of automated sample preparation systems. United States Food and Drug Administration (FDA) has also recommended procedures for validation of computer software and hardware that are used for systems that collect, report, and store raw data, such as data acquisition systems, and LIMS. Like GALP, those regulatory policies do not refer to validations performed on automated laboratory robotic systems and workstations used for sample preparation.

Automated sample preparation systems are not typically used to handle, store, or archive raw data. The typical function of these systems is to perform sample preparation and introduce prepared samples to analytical instruments that generate raw data which are collected in data acquisition systems. In some instances, data may be input, output, or handled by an automated system during sample preparation by certain equipment, such as analytical balances or barcode readers. These situations may require validations like those described for computer systems, which are beyond the scope of this chapter. In all situations, it is prudent to ensure that these data are precisely and accurately collected, transferred, and reported.

### 8.2.1 Code of Federal Regulations for Laboratory Equipment

The Code of Federal Regulations (CFR) gives specific guidelines for equipment operated in laboratories that are required to follow cGMPs and GLPs. The Code of Federal Regulations for cGMP laboratories states that: "Automatic, mechanical, or other types of equipment, including computers, or related systems that will perform a function satisfactorily, may be used in the manufacture, processing, packaging, and holding of a drug product. If such equipment is so used, it shall be routinely calibrated, inspected, or checked according to a written program designed to assure proper performance. Written records of those calibration checks and inspections shall be maintained" [12]. This CFR section also states that change control be secure and documentation be thorough.

The CFR for non-clinical GLP laboratories states that: "Equipment used for the generation, measurement, or assessment of data shall be adequately tested, calibrated, and/or standardized" [13]. This section further recommends that the testing, calibration, standardization, and maintenance procedures be specified in standard operating procedures and that thorough written records be kept for all occurrences. Laboratories that perform analyses to support clinical new chemical entity research and development typically also follow these guidelines.

In short, procedures for maintaining and testing, standardizing, and/or calibrating analytical equipment must be written down, regularly followed, and thoroughly documented upon completion. Manual and automated laboratory equipment do not differ in these requirements.

### 8.2.2 Modular and Holistic Validations

The concepts of modular and holistic validations were recently defined to describe validation requirements for computerized LC systems in GLP laboratories [12]. The authors, principally from the US FDA Division of Drug Analysis (DDA), describe validation requirements for LC systems that are integrated with computer-based controllers for which users cannot obtain complete software code or documentation. Although computerized LC systems do not chiefly perform automated sample preparation, the scenario they describe may be extrapolated to and applied to robotic systems and automated workstations that perform sample preparation.

In modular validations each system component is separately calibrated and validated, while in holistic validations the performance of all system components working together to execute an analytical method is validated. In their critique of the modular approach, the authors recognize that modular validations do not assure reliability of analytical results. Holistic validations are particularly valuable when software code and documentation are not available from manufacturers, as is often the case with automated sample preparation systems.

### 8.2.3 FDA GLP Standard Operating Procedures

The current GLP Standard Operating Procedures (SOPs) from US FDA DDA laboratories give detailed descriptions of the procedures used at their site to calibrate and validate analytical equipment and methods. The US FDA DDA SOP entitled "Maintenance and Calibration of Apparatus" specifically treats analytical instruments and computerized data handling equipment differently. In that SOP, the procedures for maintenance and calibration of analytical instruments are the same whether those instruments are "automatic, mechanical, or electronic." The SOP clearly distinguishes calibration and maintenance procedures for analytical instruments from procedures for "calculators, microprocessors, computers, and other equipment used to calculate or access data", by stating that those computerized equipment are specifically exempted from it [14].

The Method Validation section of the US FDA GLP SOPs refers to automated and manual method development and validation in the same context. Both manual and automated methods undergo the same validation procedures. These validation procedures include assessments of linearity, carry-over, precision, accuracy, recovery, and other common aspects of analytical method validations. Reference to computer testing is made only in terms of assessing the quality of data output from the analysis of standard solutions [15].

The calibration and validation procedures currently employed in automated GLP laboratories generally follow practices that are like those described in the US FDA DDA SOPs. Furthermore, many scientists in these laboratories have developed their automated methods in ways that they provide additional quality, control, and documentation that are not feasible to implement in manual systems. These value-added features are described in sections to follow.

### 8.2.4 Current Industry Practices

Most pharmaceutical GLP laboratories that have published their practices validate automated chemistry and calibrate, test, or standardize automated equipment using the same procedures and acceptance criteria that they use for manually operated methods and equipment. These procedures generally conform to a combination of the modular and holistic validation approaches described above. The overall performance of automated methods are validated and certain specific system components are regularly maintained and calibrated according to SOPs.

Some laboratories, primarily cGMP-regulated ones, validate their automated sample preparation systems using the modular approach along with software and hardware validations much like those that are used for automated data handling systems [17][18][19]. Procedures for performing this type of validation in cGMP laboratories have been well described [20-23]. The principle reasons for using computer validation procedures have been described as a means of ensuring precision and accuracy of results, and as a means preventing unauthorized or inadvertent changes and systematic errors [20].

Thorough method validations, in conjunction with certain specific instrument calibrations and regular analysis of standards and controls, function to protect manually-operated methods from random and systematic changes. These tools also function to protect automated methods from change. Additionally, automated systems can be designed to provide extensive, detailed information about system performance during validations and during routine operation.

## 8.3 Calibration of Automated System Components

Calibration is a procedure that is designed to verify and control the performance of an instrument [15, 24]. The same types of equipment that require calibration when used manually should also be calibrated when used by a robot or workstation; these are typically the types of equipment that generate raw data. Examples of equipment that require calibration include, but are not limited to, analytical balances, thermometers, pipettors, pH meters, and titrators. Components should be calibrated and documented according to the same local regulatory procedures and schedules that are used for manually-operated equipment.

The calibration status of an instrument is first verified using a traceable standard. If the instrument fails to meet specific criteria for acceptable operation, it undergoes maintenance. The instrument is not used until it meets the established criteria. Verification of the calibration status of an instrument can be automated; maintenance is typically a manual procedure.

Calibration may be easily performed at specific time intervals and it is not unusual for calibration verifications to be conducted more frequently when they are automated. For instance, it is often convenient and effortless for a calibration verification to occur at the start of each automated system run. Successful verification is commonly set as a condition of system start-up and the automated system will not run unless the criteria for acceptance are met.

Actual calibration procedures and schedules are specific to instruments and applications. Little translation is necessary to adapt calibration practices for manual equipment to equipment on automated systems. Implementation of those procedures may be different and some specific, common examples are discussed in sections below.

The piece of equipment on a robotic system or automated workstation that does not have a direct mechanical counterpart in manually-operated analyses is the automated arm or manipulator. Automated arms or manipulators should undergo regular maintenance and testing with documentation of procedures and findings. Maintenance typically comprises cleaning and oiling plus any other preventative maintenance practices that are recommended by specific manufacturers. Testing typically comprises inspection and adjustment of positions. Calibration of positions may or may not be recommended, depending on the type of system being operated. Relevant automation manufacturers should be contacted for specific recommendations.

### 8.3.1 Analytical Balance Calibration

Analytical balances are commonly included in automated system configurations because they can provide tremendous benefits when used to measure or verify masses of liquids or solids transferred by a robot during an automated procedure. Balances are also important tools to facilitate on-going calibration and validation of other system components, such as pipetting and liquid dispensing. The acceptance limits for calibration of an analytical balance on a robotic system are identical to those used to calibrate manually operated balances. Calibrations are performed at regular intervals and the results of the calibrations are thoroughly documented.

#### 8.3.1.1 Validation of Balance Data Transfer

If data from the balance will be used to make analytical calculations or quantitative decisions, the precision and accuracy of transfer of data from the balance to the robotic system controller should be validated. Accuracy of data transfer may be verified by comparing values read directly from the balance by the user with values read and reported by the system controller. Precision may be measured by verifying that the transfer of balance data is reproducible. Both precision and accuracy should be verified at masses that cover the entire applicable range of operation. The results of the manual vs. robotic readings should meet predetermined SOP requirements concerning numbers of decimal places and rounding, and the results should be thoroughly documented according to local practices.

After data transfer is validated, reliability of data transfer to the automated system controller should be regularly verified using an automated program that measures the masses of known calibrated weights and compares the resultant mass values with expected values. The values should agree with one another to a predetermined number of decimal places and the results documented. This routine test procedure may serve as a change control mechanism.

#### 8.3.1.2 Automated Balance Calibration Procedures

Balance calibration may require system hardware modifications to ensure reliable handling of calibration weights. Options include obtaining calibration weights that are designed to be manipulated by the robot or developing a robotic gripper or similar device that is designed to place standard calibration weights on the balance pan. Both of these options require that a suitable holder be affixed inside the balance to reliably receive the calibration weights.

Once the precision and accuracy of data transfer from the balance to the robotic system are verified and any necessary hardware or software modifications are made, automation of the calibration process can proceed. If data transfer is not verified or the hardware and software are not fully compatible with the system, then calibration should be performed manually.

Most balances can be calibrated in two ways: either internally, using the balance's internal calibration mechanism, or externally, using a standard, traceable weight placed directly on the balance pan. It is recommended that external calibration rather than internal calibration be automated. Automation of an internal calibration procedure typically requires validation of two-way communication between the balance and the robotic system controller. This process entails complex hardware and software interfacing tools and, therefore, requires computer validation procedures (see Chapter 12) that are beyond the scope of this chapter. The robot may be programmed to perform the external calibration in a manner that is very similar to that which is used to manually perform calibration: 1) place calibration weights on the balance pan or onto a holder designed to support the weights; 2) verify that the value it reads from the balance meets predetermined criteria for acceptance; and 3) report the calibration results to the user.

Automated calibration is often set to occur each time an automated system is started. In those cases, verification that the balance meets criteria for acceptable calibration should be made a condition of system start-up. A pass/fail mechanism may be included so that the system will not run if calibration fails to meet acceptance criteria and a report documenting these results may be automatically generated.

### 8.3.2 Calibration of Pipettors and Liquid Dispensers

The precision and accuracy of volumetric liquid dispensers and pipettors, whether manual or automated, are typically verified using gravimetric methods. That is, their calibrations are performed by measuring the mass of a standard, usually water. If verification fails to meet the established criteria for acceptance (1 mL water volume = 1 gram mass), then maintenance must be performed and calibration verification is repeated.

#### 8.3.2.1 Fully-Automated Pipettor and Liquid Dispenser Calibration

Full automation of liquid dispensing and pipetting calibration requires that a calibrated balance be included in the automated system configuration. The robotic system may be programmed to dispense multiple aliquots of water in incremented volumes that span the entire range of the liquid dispenser or pipettor that must be verified. Water is dispensed into a tared container on a balance, and the mass of each delivery is recorded by the system. An example of a common range and scheme for verifying a 1 mL (maximum) pipettor or dispenser is: 0.1 mL to 1.0 mL total range in 0.1 mL increments with six replicates (n=6) at each increment. The mass of each liquid delivery must be documented and documentation may be automated.

If verification fails to meet calibration tolerance limits, the system may be set to stop and to notify the operator of the specific error. Further control requires manual instrument adjustment by qualified personnel.

#### 8.3.2.2 Semi-Automated Pipettor and Liquid Dispenser Calibration

If a balance cannot be included on the automated system, liquid delivery should be gravimetrically verified but cannot be fully automated. This may entail manually taring appropriate containers, putting them in a rack on the robot system, then requesting the robot to dispense or pipet aliquots of specified volumes of water into the tared containers. The containers with the aliquots of water are then manually weighed to determine mass of each volume transferred and to calculate precision and accuracy [9].

#### 8.3.2.3 Calibration of Multi-Channel Pipettors and Liquid Dispensers

Multi-channel liquid dispensers or pipettors should be operated as single-channel units to be gravimetrically verified. It is not feasible to simultaneously verify multiple liquid deliveries, such as an entire row of wells on a microtiter plate, using gravimetric methods. Containers such as 96-well or 384-well microtiter plates into which multiple liquid aliquots are simultaneously dispensed can use spectrophotometric absorbance or radiometric methods to measure precision and linearity of liquid dispensing and pipetting [26]. Absolute accuracy of dispensing is most often measured gravimetrically.

## 8.4 Validation of Automated Systems

Once any necessary calibrations are performed and any validations of system components that are required have been completed, the method application may be validated. As discussed in sections above, the validation of an automated method has all the requirements of validations of manual methods. Validation requirements for specific applications are discussed in other chapters of this book and those requirements may be directly applied to automated systems.

Automated systems can provide unique benefits to help ensure thorough, successful method validations. They can help minimize analytical variability, can provide extensive monitoring and control to ensure reliability, and can also report valuable information to support validations and diagnose problems. Automation can also present unique problems that can impede success during method validation and during routine operation. The sections to follow describe some common benefits that automation brings to method validation, plus some common automation-specific concerns that should be kept in mind during method validation.

### 8.4.1 Validation of Pipetting and Liquid Dispensing

Successful calibration of an automated system pipettor or liquid dispenser does not ensure that the reagents dispensed during an application will be dispensed properly unless all reagents behave like water. When pipetting is performed manually, scientists automatically makes adjustments in their pipetting techniques to compensate for changes in a liquid's behavior. Precise and accurate dispensing and pipetting of liquids, particularly very viscous or volatile ones, on automated systems can require meticulous adjustment of many functional attributes. These may include operational variables such as liquid delivery line purges, syringe primes, pipettor tip prewets, aspiration and dispensing speed adjustments, and pre- and post-aliquot air aspirations, or hardware variables such as materials compatibility, liquid line lengths, or syringe sizes. It is therefore important to evaluate precision and accuracy of pipetting and delivery of liquids during or before validation.

It is important to remember to factor densities with the gravimetric results to accurately calculate the volumes of the liquids pipetted and dispensed.

### 8.4.2 Stability

Because one of the principle goals of automation is to increase analytical throughput, it is not unusual for automated analytical methods to continuously run for much longer durations than manual methods. Samples and reagents are in use and exposed to the laboratory environment for longer periods of time, which increases the likelihood of analytical stability problems. Common instability causative agents include temperature, humidity, and light, and their potential effects should be evaluated before or during system validation as thoroughly as possible.

Evaluation of stability typically requires that each type of sample introduced to the system, each type of reagent used by the system, and each critical reaction step be tested for integrity in the ranges of conditions it will encounter during all hours of routine operation. It is important to consider potential durations of system failures when determining the duration required for thorough stability evaluation.

Automated systems can provide extensive monitoring and control for conditions affecting stability such as temperatures, liquid levels, and gas pressures. They may also allow operation at environmental extremes and under conditions of exposure that are untenable or undesirable to humans, such as complete darkness, constant low temperature, or in the presence of radioactive, corrosive, and biological hazards.

### 8.4.2.1 Effects of Temperature

Laboratories where scientists typically work "nine-to-five" often have diurnally-cycled heating, air conditioning, and ventilation systems. When a robotic system is installed in this type of facility and programmed to run at off-hours, the daily temperature fluctuations can cause a variety of physical and analytical problems, including chemical or biological instability, condensation, gas flow changes, or fluctuating robot positions. It is prudent to consider the off-hour and weekend laboratory climate when planning and validating an automated method. If a method is expected to run over night and during weekends during routine operation, then runs during those hours should be included during validation to ensure full exposure to the range in temperature.

Known or suspected problems with temperature stability may be addressed in a number of ways. Temperature is frequently controlled using temperature-controlled storage racks, storage containers, and reaction vessels, all of which are common features on automated systems. If an entire analytical method requires meticulously controlled temperatures, then the entire automated system may be set up in a temperature-controlled room. If the room temperature is extremely high or low, the functionality of the system hardware at the required temperature should be thoroughly evaluated.

### 8.4.2.2 Effects of Humidity

Problems with humidity are primarily regional and seasonal. The potentially menacing effects that changes in humidity can have on a system may be inadvertently overlooked if a system is developed and validated during dry months of the year or in dry climates then moved to humid ones. Humidity problems are typically solved by controlling air conditioning and ventilation. Because it is difficult to simulate the effects of summer humidity during a winter validation, it may be prudent to anticipate potential problems by carefully controlling laboratory room temperature year-round.

High humidity can cause condensation on some components which can subsequently cause a number of problems with chemistry and hardware integrity. A very common problem occurs when test tubes are stored in cold racks, are removed from the rack by a robotic gripper, then are handled at room temperature. Water condenses on the test tube, then that condensation is deposited wherever the test tube is placed. The gripper and other system components that come in contact with those tubes become slippery, ultimately resulting in one of a myriad of system failures after many test tubes have been handled in this manner. The higher the humidity, the worse this problem becomes.

The effects of humidity can affect procedures that require handling of powders which can become tacky and agglutinate in humid environments. Humidity must also be carefully controlled if samples, reagents, or reactions are affected by the presence of water or by changes in atmospheric water concentration.

#### 8.4.2.3 Effects of Light

Analytical reagents and compounds may be exposed to various forms of light for longer periods on an automated system than during manual method operation. Obvious ways to control problems with exposure to light include using amber glassware or light-impenetrable secondary containers around vessels and racks. Automated systems may also be run in complete darkness or under filtered lighting.

### 8.4.3 Cross-Contamination

Cross-contamination, also known as carry-over, is a common problem on automated systems that should be thoroughly evaluated during validation. Problems with cross-contamination typically arise from serial exposure of samples to common devices or instruments, such as liquid transfer cannulae or LC injectors.

Cross-contamination may be detected during validation by alternating samples that contain maximum levels of analytes and analyte matrix with blank samples or other appropriate controls in the processing sequence. Examination of the output from analysis of the blanks or controls indicates the presence or absence of cross-contamination. Blanks and controls should be distributed throughout a run containing the maximum number of samples to be encountered during typical system operation.

If cross-contamination is found, the course of action will depend on the hardware or software involved. Apparati that have self-rinsing capabilities may simply require a slower rinse, a larger rinse volume, and/or a chemically stronger rinse solution. The common device may also be coated with non-stick material to help facilitate rinsing. In some situations, the hardware involved in the contamination problem may need to be completely replaced with an apparatus that uses ejectable, disposable, or non-stick components. A frequent example of this is the replacement of common-cannula liquid transfer hardware for pipetting hardware that uses disposable pipettor tips. Use of pipetting rather than a common cannula is very important in situations in which the residue carried over is sticky, the analytical method is extremely sensitive, or the method has a very wide analytical range.

The materials used in the engineering of automated apparati may themselves cause carry-over contamination. Compatibility of system hardware with the chemistry being automated must be taken into consideration to ensure success. Chemically inert components should be considered for applications that use common devices and/or very reactive reagents.

### 8.4.4 Method Precision and Accuracy

When properly implemented, automated systems typically yield very reproducible analytical results, as measured by within-day and inter-day analytical precision. Automated systems also exhibit little "inter-analyst" variability; therefore, analytical methods may be more easily transferred between laboratories, particularly if the systems have identical hardware and the system operators are experienced.

An overwhelming majority of scientists have used comparisons between the precision and accuracy of manual methods and those of automated methods to validate their automated systems. Comparisons are particularly valuable in situations where automation technology is new and has replaced a manual procedure. Comparisons have been performed for content uniformity and tablet assays [28-30], analysis of drugs in biological fluids [31, 32], dissolution testing [33, 34], and assays using microtiter plates [35]. Some scientists have performed statistical tests of significance on these results, using the manual method as a benchmark to detect and measure potential differences between automated and manual methods [29, 30, 36, 37]. In the cases cited, there was no significant difference between the manual and automated methods.

In situations where no manual method exists, scientists cannot compare manual assay results to automated ones. These laboratories typically develop analytical methods directly on the robotic systems and validate their automated processes by verifying that results generated by those systems meet clearly defined and well-controlled criteria for acceptance of analytical results [38, 39]. Other laboratories with well-established automation may also compare results between different automated systems [40, 41].

## 8.5 Monitoring Method Performance

The principle goals of validation are to ensure that analytical results are correct and consistent. Well-designed, thorough validations of chemistry and on-going monitoring of analytical results using controls and standards are processes that are specifically designed to monitor the reliability of analytical processes. Properly designed and regularly conducted analytical validations and equipment calibrations ensure precise, accurate, reliable results. This concept applies to automated sample preparation methods in the same way that it does to manual methods.

Automated systems are uniquely capable of supplying information that can be used to monitor system and method performance and ensure analytical reliability. Gravimetric verification and documentation of solids and liquids transferred during a method, described in sections above, is an excellent way to monitor analytical precision and accuracy and to diagnose potential problems. Errors in liquid delivery can be immediately detected and corrected when gravimetry is properly implemented. Vision systems and optical tools such as barcoding are commonly used to read, monitor, and document sample labels, liquid levels, robotic positions, or a number of other application-specific elements. Various electronic feedback mechanisms may be used to monitor the system's environment, such as the status of gas pressures, power, and peripheral instruments.

Quality control charts are used in some automated laboratories to monitor variability in precision and accuracy of analytical methods. These charts can be excellent tools for detecting subtle trends in analytical results and diagnosing analytical deviations before serious problems emerge. In one instance, analytical results from the analysis of control samples during assay validation were entered into the control chart as the established benchmark. All subsequent analyses were compared to the benchmark, thus providing on-going feedback about assay performance [42]. In another instance, control charts were used to monitor specific steps of a robotic process, such as pipetting and weighing [25].

The tools described above are valuable for monitoring many aspects of an automated system's analytical processes; however, problems that are quite difficult to diagnose nonetheless occur. When they do occur, it can be important to watch the system run through as many samples as necessary to observe all aspects of the procedure. An alternative to this tedious task is to use a video camera to record an analytical run. Videotaping can be an excellent diagnostic and trouble-shooting tool. During playback, a video recording may be fast-forwarded to selected intervals to minimize the amount of time spent looking for operations that need to be adjusted.

## 8.6 System Change Control

Clearly, it is important to keep track of and document changes to automated systems just as scientists document changes in work they perform manually. Records should be kept of the software and hardware used during specific projects or during a specific time frame. Systems with open software allow annotations to be written, dated, and initialed directly in the text by appropriate personnel. Software may be stored in electronic and/or printed form and should be stored in a location that is safe from chemical hazards and electromagnetic radiation . Hardware configurations may be stored as drawings, lists of components, photographs, and videotapes. These records should be documented and archived in a manner and according to a schedule that is appropriate to the needs and practices of the laboratory and the analytical application. Instrument logbooks are commonly used to document, store, and archive these records.

To enforce change control, some cGMP-regulated laboratories have devised elaborate sign-off systems to control and document robotic system software changes [27, 28]. Such rigorous change control measures may not be necessary in laboratories where thorough on-going analytical change control mechanisms are applied. In these laboratories, methods and operators are validated and periodically re-validated, instruments are regularly calibrated, and standards and controls are included in each automated analytical run. If significant changes in analytical methods are made, then those analytical methods are re-validated. Analytical runs are carefully designed to monitor and ensure reliability of results and analytical run acceptance criteria are explicitly documented. Such control mechanisms are implemented to protect all analytical methods from change and to ensure reliability of analytical data. Those change control mechanisms are effective when applied to automated methods just as they are when applied to manual methods.

## 8.7 Summary

Current existing FDA regulations treat calibration and validation practices for automated sample preparation instruments and methods no different from those used for manually operated instruments and methods. While it is clear and logical that automated systems and their components must fulfill the same calibration and validation requirements as manual equipment and methods, it is also clear and logical that certain aspects of automated system validations require special attention to ensure and verify reliability of analytical results. Regulations do not specifically address issues that are unique to automated system validations. However, a thread of consistency is woven through the practices of pharmaceutical laboratories and through existing FDA practices and regulations from which logical conclusions may be drawn to address the less well-defined aspects of automated system validation.

## References

1. V. Cerda and G. Ramis, *An Introduction to Laboratory Automation*, pp. 321 John Wiley & Sons, New York (1990)

2. U. Rembold, *Robot Technology and Applications*. Marcel Dekker, Inc., New York (1990)

3. M. Valcarcel and M.D. Luque de Castro, *Automatic Methods of Analysis*. Elsevier, Amsterdam (1988)

4. R.A. Felder, J.C. Boyd, J. Savory, K. Margrey, A. Martinez, D. Vaughn, in *Perspectives on Clinical Laboratory Automation* (M.S. Lifshitz and R.P. DeCresce, Eds), pp 699-711. W.B. Sauders Co., Philadelphia (1988)

5. J.J. Tomlinson, *J. Aut. Chem.* **14,** 47-50 (1991)

6. S.D. Hamilton in *Advances in Laboratory Automation-Robotics* (J.R. Strimaitis and J. N. Little, Eds.), pp 91-108. Zymark Corp., Hopkinton, MA (1991)

7. M. Hammer and J. Champy, *Reengineering the Corporation,* 233pp. HarperBusiness, New York, NY (1993)

8. S. Conder in *Proceedings International Symposium on Laboratory Automation and Robotics, 1991* (J.N. Little, C. O'Neil, and J.R. Strimaitis, Eds.), pp 116-136. Zymark Corp., Hopkinton, MA (1992)

9. E. Muttoni in *Advances in Laboratory Automation-Robotics* (J.R. Strimaitis and J. N. Little, Eds.), pp 316-323. Zymark Corp., Hopkinton, MA (1991)

10. T.R. Smith in *Advances in Laboratory Automation-Robotics* (J.R. Strimaitis and J. N. Little, Eds.), pp 139-150. Zymark Corp., Hopkinton, MA (1991)

11. United States Environmental Protection Agency, *Good Automated Laboratory Practices,* pp 233. Office of Information Resources Management, USEPA, Research Triangle Park, NC (1990)

12. W.B. Furman, T.P. Layloff, and R.F. Tetzlaff, *Journal of AOAC International.* **77**(5):1314-1318 (1994)

13. United States Food & Drug Administration, *Code of Federal Regulations 21.* Part 211.68, (1991)

14. United States Food & Drug Administration, *Code of Federal Regulations 21.* Parts 58.63 and 58.81, (1991)

15. United States Food & Drug Administration, *Good Laboratory Practices Manual*, (S.J. Logan, Ed.), pp 8.0-1. U.S. Dept. of Commerce, NTIS, Springfield, VA (1993)

16. United States Food & Drug Administration, *Good Laboratory Practices Manual*, (S.J. Logan, Ed.), pp 9.0-5-9.0-20. U.S. Dept. of Commerce, NTIS, Springfield, VA (1993)

17. K.R. Lung, J.S. Green, P.K. Hovsepian, and J.A. Short, in *Proceedings International Symposium on Laboratory Automation and Robotics, 1992* (J.N. Little, C. O'Neil, and J.R. Strimaitis, Eds.), pp 285-298. Zymark Corp., Hopkinton, MA (1992)

18. S. Scypinski, T. Sadlowski, and L. Nelson, *PI Quality*. **4**, 39 (1994)

19. E.A. Mularz in *Proceedings International Symposium on Laboratory Automation and Robotics, 1993* (J.N. Little, C. O'Neil, and J.R. Strimaitis, Eds.), pp 96-110. Zymark Corp., Hopkinton, MA (1993)

20. P.J. Motise in *Proceedings International Symposium on Laboratory Automation and Robotics, 1994* (J.N. Little, C. O'Neil, and J.R. Strimaitis, Eds.). pp 321-3 Zymark Corp., Hopkinton, MA (1994)

21. R.F. Tetzlaff, *Pharm. Tech.* **16**(3), 112-124 (1992)

22. R.F. Tetzlaff, *Pharm. Tech.* **16**(4), 60-72 (1992)

23. R.F. Tetzlaff, *Pharm. Tech.* **16**(5), 70-82 (1992)

24. N.J. Dent in *Implementing International Good Practices* (N.J. Dent, Ed.), pp153-165. Interpharm Press, Inc., Buffalo Grove, IL (1993)

25. M.J. Monahan and T.J. Giampaglia, in *Proceedings International Symposium on Laboratory Automation and Robotics, 1993* (J.N. Little, C. O'Neil, and J.R. Strimaitis, Eds.), pp 299-313. Zymark Corp., Hopkinton, MA (1993)

26. B. Feld, K.A. Perri, J.R. Mezzatesta, in *Proceedings International Symposium on Laboratory Automation and Robotics, 1993* (J.N. Little, C. O'Neil, and J.R. Strimaitis, Eds.), pp 356-375. Zymark Corp., Hopkinton, MA (1993)

27. E. Mularz, in *Proceedings International Symposium on Laboratory Automation and Robotics, 1993* (J.N. Little, C. O'Neil, and J.R. Strimaitis, Eds.), pp 96-110. Zymark Corp., Hopkinton, MA (1993)

28. W. Haller, E. Halloran, J. Habarta and W. Mason, in *Advances in Laboratory Automation-Robotics* (J.R. Strimaitis and G.L. Hawk, Eds.), pp 557-570. Zymark Corp., Hopkinton, MA (1987)

29. G.W. Inman and D.D. Elks, in *Advances in Laboratory Automation-Robotics* (J.R. Strimaitis and G. L. Hawk, Eds.), pp 689-700. Zymark Corp., Hopkinton, MA (1985)

30. LaBrecque, ME and RG Anstey, in *Proceedings International Symposium on Laboratory Automation and Robotics, 1993* (J.N. Little, C. O'Neil, and J.R. Strimaitis, Eds.), pp 111-124. Zymark Corp., Hopkinton, MA (1993)

31. L. Brunner and R.C. Luders in *Advances in Laboratory Automation-Robotics* (J.R. Strimaitis and G. L. Hawk, Eds.), pp 413-432. Zymark Corp., Hopkinton, MA (1989)

32. G.J.-L. Lee, K. Hama, and I.J. Massey, in *Advances in Laboratory Automation-Robotics* (J.R. Strimaitis and J. N. Little, Eds.), pp 545-565. Zymark Corp., Hopkinton, MA (1991)

33. L.J. Kostek, B.A. Brown, and J.E. Curley, in *Advances in Laboratory Automation-Robotics* (J.R. Strimaitis and G. L. Hawk, Eds.), pp 701-720. Zymark Corp., Hopkinton, MA (1985)

34. J.P. McCarthy, in *Advances in Laboratory Automation-Robotics* (J.R. Strimaitis and G. L. Hawk, Eds.), pp55-70. Zymark Corp., Hopkinton, MA (1988)

35. B. Feld, K.A. Perri, J.R. Mezzatesta, in *Proceedings International Symposium on Laboratory Automation and Robotics, 1993* (J.N. Little, C. O'Neil, and J.R. Strimaitis, Eds.), pp 356-375. Zymark Corp., Hopkinton, MA (1993)

36. S. Scypinski, T. Sadlowski, and P.F. George, in *Proceedings International Symposium on Laboratory Automation and Robotics, 1992* (J.N. Little, C. O'Neil, and J.R. Strimaitis, Eds.), pp 273-284. Zymark Corp., Hopkinton, MA (1992)

37. G.F. Plummer, in *Advances in Laboratory Automation-Robotics* (J.R. Strimaitis and G.L. Hawk, Eds.), pp 47-69. Zymark Corp., Hopkinton, MA (1987)

38. G. Schoenhardt, R. Schmidt, L. Kosobud, and K. Smykowski, in *Advances in Laboratory Automation-Robotics* (G. L. Hawk, and J.R. Strimaitis Eds.), pp 61-70. Zymark Corp., Hopkinton, MA (1984)

39. J.J. Tomlinson, in *Proceedings International Symposium on Laboratory Automation and Robotics, 1992* (J.N. Little, C. O'Neil, and J.R. Strimaitis, Eds.), pp 215-231. Zymark Corp., Hopkinton, MA (1993)

40. M.E. Arnold, S.J. Dennell, and A.I. Cohen in *Proceedings International Symposium on Laboratory Automation and Robotics, 1992* (J.N. Little, C. O'Neil, and J.R. Strimaitis, Eds.), pp 232-245. Zymark Corp., Hopkinton, MA (1992)

41. L.J. Kostek, B.A. Brown, L.C. Erhart, and J.E. Curley, in *Advances in Laboratory Automation-Robotics* (J.R. Strimaitis and G. L. Hawk, Eds.), pp 311-328. Zymark Corp., Hopkinton, MA (1989)

42. T.L. Lloyd, T.B. Perschy, A.E. Gooding, and J.J. Tomlinson, *Biomed. Chromatogr.* **6**, 311-316 (1992)

# Chapter 9

# Biotechnology Products

*G. Susan Srivatsa*

## 9.1 Introduction to Biopharmaceuticals

Using the broadest definitions, a biopharmaceutical may be defined as any molecule intended for therapeutic, prophylactic or diagnostic use that was generated by a manufacturing process incorporating the use of living cells. This definition applies even if the therapeutic entity is thought to be identical in molecular structure to a naturally occurring substance. The most common biopharmaceuticals to date are proteins derived from recombinant DNA technologies and monoclonal antibodies resulting from hybridoma technology.

In simple terms, recombinant DNA technology is the manipulation of nucleic acids (DNA, RNA) in a host cell to produce a new protein sequence or amplify production of an endogenous protein considered to be of diagnostic or therapeutic value. The experimental conditions are optimized such that the desired protein is over-expressed in the host cell system of either prokaryotic or eukaryotic origin to attain high yields of the product. In hybridoma technology, an antibody producing B-lymphocyte is fused with a myeloma cell and the resulting hybridoma cells cloned. The purified product of a cloned hybridoma cell line is a monoclonal antibody directed at a specific antigenic determinant.

Advances in molecular biology, genetic engineering, process purifications, analytical chemistry and related disciplines have led to the production of large quantities of highly purified proteins facilitating drug development efforts. A number of these proteins such as insulin, human growth hormone (HGH), α-interferon, γ-interferon, tissue plasminogen activator (TPA), granulocyte colony stimulating factor (G-CSF), granulocyte macrophage colony stimulating factor (GM-CSF), Interleukin-2, erythropoietin, hepatitis B-vaccine and murine monoclonal antibody have already been approved by the Food and Drug Administration (FDA) for human use in the United States while a number of others are currently under active development as potential therapeutic agents.

## 9.2 Regulatory Requirements of Biopharmaceuticals

In order for a drug to be approved for human use it is the responsibility of the manufacturer to demonstrate its safety and efficacy by means of well controlled preclinical animal studies and human clinical trials [1, 2]. Once the inherent safety and efficacy of the intended pharmaceutical has been established, the manufacturer is then required to demonstrate the capability to consistently manufacture the intended drug product of a quality that is at the very minimum equivalent or preferably better than that used in the safety assessment and clinical evaluation phases of development. This necessitates the availability of methods that are capable of assessing the quality of the safety-assessment supplies and the clinical supplies. The potential molecular complexity of biopharmaceuticals has created an unprecedented opportunity for the advancement of analytical biotechnology as pharmaceutical analytical chemists strive to develop appropriate methodologies that can adequately characterize these molecules. This situation has facilitated improvement of some classical biochemical techniques such as slab gel electrophoresis and accelerated development of others such as tryptic mapping, carbohydrate analysis, bioassays, ELISA, mass spectrometry and capillary electrophoresis to meet the regulatory needs for drug approval.

The marketing of all pharmaceuticals in the United States is regulated by the Code of Federal Regulations for Drugs, 21 CFR, Chapter 1, Sub-chapters C and D. Biological products which include all vaccines, antibiotics, hormones, human blood or blood-derived products, immunoglobulin products, products containing intact cells, fungi, viruses or virus pseudotypes, proteins produced by cell culture or transgenic animals and animal venoms; synthetically produced allergenic products; and drugs used for bloodbanking and transfusion are further regulated by the Code of Federal Regulations for Biologics, 21 CFR, Chapter 1, Subchapter F. While the fundamental issues regarding safety and efficacy of biopharmaceuticals are essentially the same as those for conventional pharmaceutical agents, the complexity of biotechnology products has necessitated additional guidance regarding their manufacture and control. This need has been met in part by the collaborative efforts of the biopharmaceutical industry, academia and the regulatory agencies, resulting in the issuance of a number of Guidelines [3-5] and "Points to Consider" documents [6-8] to aid the industry in assuring the identity, purity, safety and potency of new drugs generated from recombinant DNA and hybridoma technologies.

More recently, there has been a major initiative to harmonize the technical requirements for the registration of pharmaceutical products in the United States, the European Union and Japan prompted by the need to streamline the drug development process, to minimize drug development costs and to accelerate drug approval. This has led to the development and issuance of a number of new guidelines as part of the tripartite harmonization initiatives by the International Conference on Harmonization (ICH) of Technical Requirements for Registration of Pharmaceuticals for Human Use. [4, 5] The six sponsors of the ICH are the US Food and Drug Administration, US Pharmaceutical Research and Manufacturers Association, the European Commission, the European Federation of Pharmaceutical Industry Associations, the Japanese Ministry of Health and Welfare, and the Japanese Pharmaceutical Manufacturers' Association. Many of the new or revised guidelines are still in the draft stages and are expected to be finalized and issued shortly. The ICH guidelines are discussed in more detail in Chapter 3 of this publication.

## 9.3 Analytical Requirements of Biopharmaceuticals

Every lot of a pharmaceutical product intended for human use must be assured of its safety, efficacy and purity. This is generally accomplished by application of a variety of techniques to meet the key analytical requirements of identity, purity, potency (biological activity), strength (concentration), safety and stability.

Minimum requirements for the identity tests are met by comparison of the property of the analyte to that of a pre-established, well-characterized reference standard. A combination of **qualitative** tests are often used to unequivocally establish the identity of a biopharmaceutical which may include molecular size/weight determination by size exclusion chromatography or gel electrophoresis, amino acid analysis, peptide mapping and protein sequencing. In addition to confirming the identity of the macromolecule these techniques also provide key information regarding the primary and secondary structure.

Purity analysis of a pharmaceutical protein is better described as impurity analysis as there is no single analytical technique that yields a purity value of a given sample. The approach is usually to identify and quantitate all known or potential impurities in the drug product from which an assessment of overall purity can be made. [9] The types of impurities that are found in protein pharmaceuticals and the analytical techniques commonly used to identify/quantitate them are shown in Table 9.1. Proteinaceous impurities may be introduced from a number of sources during the synthesis, isolation and purification phases of the manufacturing process. These include contaminating host cell proteins which may be copurified with the product; altered or inactive forms of the protein; products of the interaction of the intended protein with reagents used in isolation and purification; proteins added for growth promotion during the fermentation stage of the process and proteins that may have leached into the final product during purification by affinity chromatography.

Process related impurities, also known as non-adventitious impurities, are substances present in the raw materials, bulk drug substance or final drug product that are not considered to be active ingredients, additives or excipients. They can be innocuous in that they do not pose a health hazard or deleterious in that they may be considered a health or safety concern, particularly with respect to toxicity, carcinogenicity or immunogenicity. Deleterious impurities must be controlled within safe limits and their levels must be determined in every lot of drug product using appropriate analytical techniques. Contaminants, sometimes referred to as adventitious impurities are biological or chemical agents that may be accidentally introduced into the drug product. While these are difficult to control entirely, such events must be anticipated and appropriate broad spectrum, non-specific analytical methods such as LC with low wavelength detection should be incorporated into the drug product monograph such that these agents, if present, will be detected. Analytical methods should be developed to detect and quantitate impurities or related compounds that are considered to be endogenous to the process and would be expected to be present in the drug.

**Table 9.1**
Impurities found in protein pharmaceuticals

| Impurity | Analytical Techniques Used |
| --- | --- |
| Host Cell Proteins | SDS-PAGE, Immunoassays |
| Monoclonal Antibodies | SDS-PAGE, Immunoassays |
| Related proteins | SDS-PAGE, Immunoassays, LC |
| Nucleic acid impurities | Hybridization assays, antibody or sandwich assays |
| Product Variants | |
| Deamidation products | IEF, ion exchange LC, MS |
| Oxidation products | LC |
| Amino acid substitutions | LC-peptide mapping, MS |
| Aggregrated forms | Size Exclusion LC, SDS-PAGE |
| Proteolytic products | SDS-PAGE, LC, IEF |
| Homologous host cell proteins | Immunoassays |
| Infectious agents | Reverse transcriptase assay, cell culture cytopathic effects |
| Endotoxin | LAL, rabbit pyrogen test |

Impurities in biopharmaceuticals (Table 9.1) can also be broadly classified as minor (0.5 % or less) and major impurities (0.5 - 1.0%). Major impurities are generally identified and attempts are made to minimize their presence. Often toxicological, pharmacological and immunological data are acquired on these isolated impurities to alleviate safety concerns. Minor impurities are usually identified and included as part of the release specification of the drug. The most common methods used for impurity analysis of biopharmaceuticals are electrophoretic techniques (SDS-PAGE, IEF, CE, etc.), chromatographic techniques (RP-LC, IEC, SEC, etc.) and immunoassays (RIA, ELISA, sandwich ELISA, etc.).

Safety concerns of biopharmaceuticals arise from the potential presence of viruses, mycoplasma, nucleic acids, infectious agents and pyrogens in the drug product [10]. Virus testing is often performed at various points during production [11]; and process validation studies are generally performed to demonstrate that the adventitious agents such as viruses and DNA are removed or inactivated by the purification process [12, 13]. Mycoplasma assays must be performed in compliance with the Code of Federal Regulations [14]. Pyrogenicity may be determined using the Rabbit Pyrogen test [15] or the Limulus Amebocyte Lysate (LAL) assay described in the USP [16] as well as in a "Points to Consider" document [17]. In addition to the above, the sterility of each lot of drug substance and drug product must be established using validated procedures [18].

Potency and strength of the drug product are determined by assay against an international reference standard of well defined potency. These assays may use live animals to record biological activity or may adopt an *in vitro* cell culture based assay or a biochemical method that has been validated to adequately model the desired pharmacological activity *in vivo*.

## 9.4 Validation of Analytical Biotechnology Techniques

The ICH guideline (see also Chapter 3) summarizing the validation requirements of analytical procedures used for the quality control of pharmaceutical products was recently issued [5]. The key elements of validation for the various types of analytical tests such as identity, impurity limits, quantitative impurity limits and purity/potency tests discussed in this guideline are shown in Table 9.2. The experimental parameters that contribute significantly to the validation of the analytical methodologies commonly employed for the characterization of biopharmaceuticals are addressed below.

**Table 9.2**
Validation elements of analytical methods used for biopharmaceuticals

| Type of Method | Identity | Impurities Test | | Potency Assay |
|---|---|---|---|---|
| | | Quantitative | Limit | |
| Accuracy | - | + | - | + |
| Precision | | | | |
| Repeatability | - | + | - | + |
| Intermediate Precision | - | + | - | + |
| Reproducibility | - | - | - | - |
| Specificity | + | + | + | + |
| Detection Limit | - | + | + | - |
| Quantitation Limit | - | + | - | - |
| Linearity | - | + | - | + |
| Range | - | + | - | + |

### 9.4.1 Mass Spectrometry

Mass spectrometry has been used extensively by pharmaceutical analytical chemists as a means of establishing the identity of drug substances by confirmation of their molecular weight. Until relatively recently difficulties in volatilization and subsequent ionization without fragmentation and/or degradation has precluded the use of this technique for accurate molecular weight determinations of biopharmaceuticals. However, recent advances in soft ionization techniques such as fast atom bombardment (FAB), matrix assisted laser desorption (MALDI), electrospray (ES), and plasma desorption ionization have made possible ionization of macromolecules without fragmentation allowing access to molecular ion information [19, 20]. Of these, electrospray appears to be the most common technique employed for the molecular weight determination of biopharmaceuticals. In this technique, a stream of an acidic solution containing the analyte is sprayed through a small orifice. A high electrical potential is applied to the sample stream resulting in the formation of highly charged droplets which are nebulized by a carrier gas to a fine spray. Upon evaporation of the solvent, the resultant multiply charged analyte molecules are introduced by electrospray into the analyzer and detected by mass analyzers such as quadrupoles. Because of multiple charges on the analytes, they are detected at much lower mass to charge ratios (m/z) than conventional mass spectrometry experiments making accurate molecular weight determination of large macromolecules feasible. Nevertheless, routine estimation of molecular weight of biopharmaceuticals as an identity test is most commonly performed using chromatographic or electrophoretic techniques.

Mass spectrometry has played a key role as an adjunct to peptide mapping studies. An essential element of validation of peptide maps is structural identification of the peptide fragments following separation by LC (for a more detailed discussion of peptide mapping, see Section 9.4.3). Detection and characterization of coeluting peaks is a key advantage of the use of on line LC-MS, which may otherwise go undetected or require lengthy isolation and characterization by amino acid analysis or other techniques. Efficiency of the LC-MS interface is critical as low level peptide fragments may be undetected by MS due to inadequate ionization.

A number of interfaces are currently available for on-line LC-MS experiments such as thermospray, continuous FAB and electrospray. Because these studies are rarely conducted as part of the routine quality control testing of a drug product but more at the research stage to aid in LC methods validation or other structural characterization studies, LC-MS techniques are not generally "validated" according to FDA guidelines. Rather, standard good laboratory practices are followed tracking instrument performance such as sensitivity and mass resolution on an ongoing basis. Incorporation of external or more preferably internal molecular weight standards of similar molecular structure should also be considered to assure accurate and reproducible mass assignments.

### 9.4.2 Amino Acid Analysis

This approach involves hydrolysis of the protein of interest down to the individual amino acids, which are then separated chromatographically and quantitated. Hydrolysis of the protein is carried out in highly acidic conditions, typically 6N HCl at greater than 100 °C for up to 24 hours. The sample is then subjected to chromatography (ion exchange or reversed phase) coupled with either pre- or post-column derivatization with ninhydrin to facilitate UV-visible detection or more recently, using phenylisothiocyanate, dansyl chloride or *o*-phthalaldehyde to allow highly sensitive fluorescent detection. The peaks in the chromatogram may be readily identified and quantitated by means of suitably derivatized amino acid standards which are commercially available. Amino acid analysis can offer a high degree of accuracy for small proteins containing less than 250 base residues. For larger proteins, incomplete recovery of certain amino acids may lead to inaccuracies that must be overcome by careful optimization of the hydrolysis reaction. A key limitation of this technique is its relative insensitivity to detect point mutations which effectively precludes its use for measuring purity or genetic stability of proteins of molecular weight larger than 5 - 10 KDaltons.

Certain amino acids such as serine and threonine are susceptible to degradation necessitating kinetic analysis to determine initial concentrations. If tryptophan is present in the protein of interest, modified hydrolysis conditions need to be assessed as complete destruction of tryptophan can occur in the relatively harsh hydrolysis conditions of 6N HCl. Cysteine and methionine often require pre-oxidation to cysteic acid and methionine sulfone, respectively, for accurate quantitation. Amino acid recoveries are known to be dependent on protein sequence often precluding the use of hydrolysis conditions that have been previously optimized for a different protein. Solution components such as buffer salts, particularly phosphates, divalent metal ions and sugars also impact amino acid recoveries. For large proteins, staying within the linear dynamic range of the detector for all amino acids may be difficult. For these reasons, unique sample preparation, hydrolysis conditions, and dilution protocols need to be optimized for every protein. Strict adherence to these well defined conditions is critical as relatively minor deviations may result in inaccuracy and poor precision between measurements. Certain enhancement in precision may be achieved by automating the entire process of digestion, derivatization, separation, detection and data processing. However, while automated amino acid analyzers can contribute to more reproducible data, delineation and control of the experimental factors surrounding the hydrolysis conditions remains a key variable in achieving accurate and reproducible amino acid analysis. Once validated, amino-acid analysis can serve as an excellent technique with the desired sensitivity, accuracy, precision, throughput and automation features to make it the method of choice for establishing the identity of a biopharmaceutical. Amino acid analysis when used in a quantitative mode can be a direct measure of protein content. This is discussed in more detail in section 9.4.8.

### 9.4.3 Peptide Mapping Analysis

Peptide mapping, often referred to as fingerprinting, involves enzymatic or chemical cleavage of a protein under well controlled conditions to yield peptide fragments, which are subsequently separated by a suitable chromatographic technique such as reversed phase LC. The resulting chromatogram is a fingerprint of the protein, which, when compared to that of an authentic reference standard, can serve as confirmatory evidence of the identity of the protein. This technique is conceptually similar to that of mass spectrometry with electron impact ionization in which a large molecule is fragmented to yield information about its molecular structure. Peptide mapping is particularly powerful as it can detect minor differences in the primary sequence of a protein that are not usually detected by the common separation techniques conducted on the intact macromolecule. This is especially important for the detection and identification of single point mutations or mistranslation of the cDNA sequence leading to single or multiple incorrect amino acids in the protein sequence [21].

The enzymatic approach to achieve selective fragmentation uses endoproteases, most commonly trypsin, for the digestion step. The chromatogram of the resulting peptide fragments is called a tryptic map [22]. The selection of a suitable enzyme is dependent upon the primary sequence of the protein, i.e. its susceptibility to fragmentation to a level that affords meaningful sequence information. Peptide mapping may be used to determine the presence and the location of disulfide bonds *via* reduction and chemical modifications of the cysteine residues or through direct analysis by electrochemical detection. Comparison of a tryptic digest with and without endoglycosidase treatment can serve to detect and identify the location of oligosaccharide side chains on glycoproteins. Sites of phosphorylation may be distinguished from the peptide maps by comparing digests of phosphorylated and dephosphorylated protein [23].

As in the case of amino acid analysis, a key factor influencing the success of peptide mapping experiments is the optimization and control of the conditions during enzymatic digestion. Ruggedness of this step of the analysis is influenced by many factors including temperature, pH, type of buffer used, purity and lot-to-lot consistency of the enzyme, length of digest, stability of the fragments and protein concentration. Low or inconsistent purity of the enzyme can lead to non-selective cleavage due to contamination by other endo and exoproteases. Care must be taken to assure that the peptide fragments are not lost during sample handling due to nonspecific adsorption to glass, filters or other surfaces, with which the sample may come in contact. The addition of surfactants may be necessary, particularly for hydrophobic proteins, to improve recoveries and consequently, reproducibility of the peptide maps. For large proteins, digestion conditions need to be optimized such that the peptide fragments generated are of a size that affords structural information while being small enough to be separated with good resolution by chromatography. This is a common problem with the use of trypsin to cleave large proteins resulting in so many fragments that effective resolution by chromatography is difficult. Also, larger number of fragments are likely to have much higher limits of detection. In such cases where the protein is greater than 250 amino acids in length, endoproteases such as tosyl-L-phenylalanine chloromethyl ketone (L-TPCK)-treated trypsin are often used to inhibit the action of chymotrypsin, a contaminant present in most trypsin preparations. Stability of the tryptic peptide fragments during analysis is important as peptides such as glutamine and asparagine are susceptible to deamidation reactions leading to glutamic acid and aspartic acid respectively. The rates of these reactions vary with experimental conditions leading to spurious product peaks that appear as a function of time. For this reason and to establish peak purity (a validation requirement for all chromatographic methods), structural characterization of each peak in the chromatogram is a necessary aspect of the validation of peptide mapping. There are many approaches to achieving this including collection of each chromatographic peak for subsequent analysis by amino acid analysis, sequencing and/or mass spectrometry. Of these, LC coupled with mass spectrometric detection has been the most favored technique for confirming the structure of peptide fragments [24-26].

To use peptide mapping as a tool for assuring the consistency of a protein pharmaceutical on a lot-by-lot basis, it is essential to determine the specificity of the method, i.e. its ability to detect and quantitate single point mutations in the sequence. This can be accomplished by spiking the protein samples with authentic single amino acid mutants and evaluating the capability of the method to separate and detect these species. Other considerations in establishing peptide mapping for routine analysis involve ascertaining the other elements of validation such as accuracy, precision and linearity of detector response which are influenced by a number of chromatographic variables such as the solvent gradient employed to separate the peptide fragments, mobile phase composition, solvent flow rate, column temperature and column stability. Once chromatographic conditions have been optimized, the impact of minor changes in common experimental parameters such as mobile phase composition, solvent gradients, flow rates on the retention times and peak areas of each peptide fragment need to be ascertained as part of the validation of the chromatographic phase of the peptide mapping analysis. Linearity of the chromatographic response should be established for the most common range of initial protein concentration. Fragments that display non-zero intercepts may indicate poor chromatographic resolution or incomplete/partial cleavage. Within-day and between-day reproducibilities should be determined individually for the digestion and chromatographic steps as well as the overall reproducibility of the peptide-mapping analysis for a given sample.

It is apparent from the above discussion that validation of a peptide mapping technique requires significant time and effort. However, once the digestion and chromatographic steps of the analysis are well established and fully validated, peptide mapping has a much better chance of detecting variations in the primary structure of a protein than any other structural measurements that could be made on the intact macromolecule.

### 9.3.4 Protein Sequencing Analysis

Protein sequencing is a classical biochemical technique that is commonly used for establishing the identity and the primary structure of a pharmaceutical protein. Although either amino terminal (also known as the N-terminal) or carboxy terminal (also known as the C-terminal) sequencing may be sufficient for routine analysis, both approaches are often utilized at least once, usually on the reference standard, to assure absolute sequence authenticity.

### a. Amino Terminal Sequencing

The most common technique, known as Edman degradation analysis [27, 28], involves the coupling reaction of the amino terminal residue of the protein with phenylisothiocyanate (PITC). The resulting phenylthiocarbamyl (PTC) amino acid derivative is cleaved from the protein by use of a perfluoridated organic acid such as trifuoroacetic acid or heptafluoroacetic acid, thereby subjecting the next amino acid in the sequence to a similar coupling/cleavage reaction. The cycle is repeated, generally 8-10 times, and the modified amino acid residues, (anilinothioazolinones, ATZ) that have been cleaved from the protein are converted to a phenylthiohydantoin-amino acid (PTH-AA) in the presence of acid and heat. The PTH-AA derivatives are subsequently analyzed by a suitable separation technique, most commonly reversed phase LC with UV or fluorescence detection. The PTH-AA derivatives in the sample are identified by comparison of their k' values with authentic PTH-AA standards. This entire process has been automated successfully and protein sequencers are readily available from a number of commercial sources. These sequencers provide a high degree of precision due to automation of key steps that normally contribute to variability between experiments. These are discussed later in this section.

A key advantage of Edman-degradation analysis is its ability to detect the presence of and obtain an approximate measure of the extent of intrachain cleavages as well as secondary N-termini within the protein. These secondary N-termini, particularly if present at tryptic sites such as arginine or lysine, are not as readily detected by any other technique. The chromatographic peaks corresponding to these cleavages, although of much lower intensity, may enable identification of the location of the alternate cleavage sites as well as the degree of heterogeneity in the sample. In addition, the use of fluorescence detection affords enhanced sensitivity allowing favorable limits of detection for most applications.

While some quantitative information may be attained by Edman-degradation analysis, it is important to note that it is primarily a qualitative technique. There are a number of limitations in using N-terminal sequencing for establishing the absolute identity and the primary sequence of a protein. Phenylthiohydantoin derivatives of serine, threonine and cysteine may be degraded by the chemistry involved in Edman sequencing thereby making their detection/quantitation problematic. Approaches to stabilize this degradation have included chemical modifications prior to Edman chemistry. In addition, glycine and proline are known to undergo slow and, therefore, incomplete conversion resulting in peaks in the chromatogram that are difficult to identify or quantitate. Nevertheless, validation of this sequencing paradigm for routine analysis of a protein is possible provided there is a good understanding, *a priori*, of the amino acid content and the unique nature of the protein of interest. Also, systematic experiments should be conducted to arrive empirically at customized derivatization and cleavage conditions for that protein. This may not necessarily mean that the undesired degradation reaction has been avoided nor that slow or incomplete reactions have been forced to completion but simply that the experimental parameters have been defined such that these unfavorable reactions occur in a consistent manner from experiment to experiment to allow meaningful interpretation of the sequencing data. Once this is accomplished, reproducibility of the technique can be quite good particularly with the advent of automated sequencers that are capable of a high degree of precision in controlling experimental conditions. This technique is also subject to the normal validation expectations of any analytical chromatographic technique particularly because of the use of shallow solvent gradients used to separate the amino-acid derivatives. These include chromatographic parameters such as system suitability, resolution, peak shape, peak purity, sensitivity and linearity of response which are discussed in more detail in Chapter 2.

### b. Carboxy-Terminal Sequencing

Sequential cleavage of amino acids from the carboxy terminus may be accomplished *via* chemical or enzymatic means to yield similar information as to protein identity and primary structure. Reaction with hydrazine or ammonium thiocyanate have been reported to yield amino-acid derivatives that may be separated, identified and quantitated by RP-LC. The enzymatic approach involves the use of natural exopeptidases such as carboxypeptidase or other enzymes capable of sequential digestion of the protein. As with the use of any naturally derived reagent, the enzymatic approach has poor precision because of lot-to-lot variability in the reagent enzyme due to presence of endopeptidases or other exopeptidases as contaminants that compete in a non-selective fashion in the digestion reaction leading to complicated reaction kinetics. There have been several approaches to improve C-terminus sequencing protocols particularly for proteins not readily amenable to Edman sequencing. One such method is the use of protease or cyanogen bromide to digest the protein of interest prior to separation and detection [29].

As with amino-terminal sequencing, validation of a C-terminal sequencing method requires systematic evaluation of the key variables that have an impact on the efficiency and reproducibility of the digestion reaction such as purity and lot-to-lot variation of the cleavage reagent, its concentration, length of digestion, temperature of the solution, effects of ionic strength and presence of other sample components. For the reasons mentioned earlier, this is inherently easier to accomplish if a chemical reagent is used instead of an enzyme due to their ready availability in high purity and good lot-to-lot consistency. Once the digestion portion of the method is well defined and deemed to be suitably accurate and reproducible, the chromatographic separation and detection system can then be validated against their respective parameters.

### 9.4.5 Electrophoretic Methods

Electrophoresis, a technique in which the molecules are separated based on their differential migration towards an electrode when placed in an electric field, represents one of the oldest methods for the separation of proteins. Various modes of electrophoresis may be used to not only confirm the identity of a protein but also to establish the purity and homogeneity of the protein in a given sample. Because of its ability to detect chemical and structural changes, electrophoretic methods have also been used successfully to monitor the stability of pharmaceutical proteins.

### a. PolyAcrylamide Gel Electrophoresis (PAGE)

Polyacrylamide gel electrophoresis (PAGE) is by far the most common mode of electrophoresis. In SDS-PAGE analysis, the protein of interest is denatured with a surfactant sodium dodecylsulfate (SDS) producing a protein-detergent complex whose net negative charge is proportional to its mass. When placed in an electric field, the protein molecules migrate towards the electrode at a rate that is proportional to their molecular weight. Following electrophoresis, the proteins in the slab gel are detected by means of direct staining techniques using either silver stains or Coomassie Brilliant Blue [30].

The identity of the protein may be confirmed by co-migration of the sample with a reference standard. However, as the extent of migration on an SDS-PAGE experiment is insufficient to establish the identity of a protein it must be used in conjunction with other methods for unambiguous identification. In the non-reducing mode, SDS-PAGE is an excellent technique for assessing the state of aggregation or oligomerization of a protein as these will appear as late migrating species in the gel provided these aggregates are stable to the denaturing conditions of SDS-PAGE. The method may also be run without detergent treatment in a "native gel" mode to evaluate the extent of aggregation although such separations tend to be of relatively poor resolution to yield reliable molecular weight information. An estimation of purity may be made by comparison of the sample with that of the reference standard to detect the presence of new impurities.

The amount of protein in the band may be determined either qualitatively using visual observation or quantitatively using laser densitometry. Although not readily amenable to laser densitometry due to high backgrounds, silver staining has much higher sensitivity allowing detection of protein bands at sub-ng levels [31]. Detection in the range of 1 ng corresponding to a concentration detection limit of 200 μg/mL can routinely be achieved for a sample load of ~ 5 μg of a typical sample. Estimation of the impurity levels requires running protein standards of increasing dilutions alongside the sample on the same gel until the intensity of the impurity band visually matches that of a known dilution of the standard. This quantitation technique is similar to that practiced for thin layer chromatography (TLC) and can be adequate for a limit test. While common proteins such bovine serum albumin (BSA) have been used as standards, accuracy may be improved by using a reference standard of the protein of interest to make such assessments. Still, the protein sample may contain several impurities below the detection limit, which may add up to a significant percentage of total impurity in the sample. Enzyme linked immuno-sorbent assays (ELISA), because of their increased sensitivity for the detection of antigenic protein impurities are recommended in combination with electrophoretic methods to achieve absolute quantitation of protein impurities in a given sample.

If a quantitative impurity test is warranted, the use of Coomassie Brilliant Blue staining with laser densitometric quantitation is preferable. In order to validate such a method for routine quantitative use, a number of variables need to be evaluated and controlled. The effectiveness of staining is highly dependent on sample loading, which can vary from protein to protein. This is a key factor in obtaining high quality gels suitable for quantitation and this parameter must be optimized for reliable analysis. Related protein impurities may not stain to the same degree as the protein of interest leading to potentially erroneous values of impurity content. In cases where the identity of the impurity is not known and/or unavailable for use as a standard, one has to assume that the staining of the impurity is similar to that of the protein of interest. In most cases, however, this is a relatively safe assumption. If Coomassie Brilliant Blue staining with laser densitometric detection does not have the desired limit of detection, then the use of other techniques such as capillary electrophoresis with laser induced fluorescence (LIF) detection may be considered.

There are several factors that impact the ruggedness of PAGE analysis. For the separation phase of the analysis, many of the factors that affect the extent of migration of the protein are not critical as the migration of the sample is usually compared with that of a reference standard of the protein of interest, usually run simultaneously on the same gel, as a confirmation of identity. Factors impacting resolution, on the other hand, are significant and must be controlled to assure adequate separation of protein impurities of similar molecular weight. Peak purities are difficult to ascertain in a slab gel experiment and could lead to an overestimation of protein purity because of co-migration of one or more related impurities with the parent peak.

An essential element of validation of SDS-PAGE analysis is evaluation of the gel under reducing conditions. Reduction of the intramolecular disulfide bonds of the protein using such reagents as β–mercaptoethanol or dithiothreitol unfolds the protein, which will then migrate at a slower rate consistent with a higher apparent molecular weight in the gel due to its increased hydrodynamic volume. Appearance of bands at lower apparent molecular weights may indicate the presence of multiple chains in the original protein, which are separated into their respective single chains under the reducing conditions of the experiment. Also, the presence of intramolecular cleavages known as "clips" can result in the same phenomenon. These new bands can be quantitated in much the same way as described earlier. Further treatment of the sample using iodoacetic acid or iodoacetamide results in alkylation of the free sulfhydryl groups leading to further structural changes detectable by SDS-PAGE. While these techniques are very effective in elucidation of secondary structure and identities of possible protein impurities, validation of the reproducibility of such procedures requires optimization and strict adherence to the experimental conditions used for the reduction and derivatization reactions. Otherwise, the gels may contain artifacts that could interfere with the separation precluding meaningful interpretation of the data. Finally, the quality of PAGE experiments are often operator dependent making validation of ruggedness difficult. For this reason, capillary electrophoresis is rapidly gaining popularity for the analysis of protein samples. A more detailed discussion follows.

### b. Isoelectric Focusing

Isoelectric focusing (IEF) is a mode of electrophoresis in which proteins are separated by their migration in an electric field over a pH gradient. The pH gradient may be formed by casting a thin layer of gel containing a large series of carrier ampholytes. In an electric field, the carrier ampholytes are arranged in the order of increasing isoelectric point (pI) from the anode to the cathode thereby generating and maintaining a local pH corresponding to their pI. Thus a uniform pH gradient is created across the gel matrix. Samples are applied to the gel surface and upon application of an electric field, each protein component migrates to the region in the gradient where the pH corresponds to its pI. At that point the protein is electrically neutral and becomes stationary or focused in the gel. The protein bands may then be visualized by staining techniques similar to those described above for PAGE analysis. Alternatively, a pH gradient can be generated in the gel when the matrix is cast through addition of charge modified monomers such as immobilines to the gel solution prior to polymerization of the gel. Thus isoelectric focusing is complementary to PAGE analysis, a separation that is primarily based on molecular size. The combination of IEF and PAGE is very powerful in elucidating the nature of protein impurities or degradation products in biopharmaceuticals.

While IEF is a powerful tool used primarily for confirming the identity of a protein biopharmaceutical, upon suitable validation it can also be used as a stability-indicating method to monitor changes to the protein over time [32, 33]. Isoelectric focusing is normally run in a native gel using wide pore polyacrylamide or agarose but the addition of nonionic detergents, nonionic chaotropic agents or reducing agents during sample preparation may serve to dissociate molecular complexes and aggregates resulting in enhanced resolution.

### c. Capillary Electrophoresis

Capillary electrophoresis (CE) has recently emerged as the technique of choice in obtaining high resolution electrophoretic separations of proteins and peptides [34]. These separations are fast, relatively easy to perform, of very high resolution, amenable to automation and may be linked to a variety of detectors. Also, a variety of electrophoresis experiments such as conventional zone electrophoresis, IEF, isotachophoresis and micellar electrokinetic chromatography (MEKC) may be conveniently performed using the same instrument with relative ease. Capillaries may be also be packed with gels to operate in the capillary gel electrophoresis mode to mimic PAGE separations but with significantly higher resolution and ease of operation. Method development using CE is generally straightforward as a number of separation variables such as different buffers, pH, temperature, and additives may be evaluated in a short time by simply flushing the appropriate solutions through a single capillary. Capillaries may be coated with various agents to allow manipulation of selectivity using wide pH ranges at below and above the isoelectric points of the protein of interest.

Capillary electrophoresis is rapidly gaining popularity for assessing the purity of proteins in part due to its ability to separate charged structural variants such as deamidation products with very high resolution. For erythropoietin (EPO) whose carbohydrate structure makes up 40% of its weight, CE was demonstrated to resolve the various glycoforms by manipulation of pH and the incorporation of organic modifiers [35]. When linked to mass spectrometry, CE is a powerful technique for structural characterization of proteins, particularly for products of peptide mapping. IEF experiments may also be conducted in the CE mode by separation of the protein in a stable pH gradient. However, IEF by CE poses special challenges in achieving good accuracy and precision, such as minimization of endosmotic flow to facilitate on line detection, mobilization of the peaks to the detector upon completion of focusing without loss of electrophoretic resolution and maintenance of a stable pH gradient across the capillary column by maintaining a uniform ampholine distribution over the pH range of interest. The IEF experiment is further complicated by the high background absorbance of ampholines at 280 nm which are manifested as unstable baselines in the high sensitivity detections employed by CE. SDS-PAGE type separations to assess molecular size can be performed in the gel mode as a high resolution, high sensitivity alternative to slab gel electrophoresis. This is accomplished by covalently bonding polymerized polyacrylamide to the fused silica capillary by means of cross linking agents. However, CGE suffers from poor reproducibility due to difficulty in obtaining uniform gels. In addition, at the high voltages commonly used in CE, the carboxylate ions are formed as a degradation product of acrylamide leading to destruction of the gel matrix.

It is clear that CE is an indispensable technique for protein analysis as evidenced by its extensive use in various facets of drug development from in process control during fermentation to stability monitoring of protein drug products. Although CE is still primarily being used in a research mode, its utility in providing reliable analysis in a quality control environment is rapidly being realized. Capillary electrophoresis has been shown to be equivalent to RP-LC for the analysis of human growth hormone in terms of its linearity of response, precision and sensitivity. In addition, use of internal reference standards have been demonstrated to significantly improve reproducibility of migration times and detector response between injections. More recently, CE was shown to have adequate accuracy, precision, selectivity, linearity of response and limit of detection to assay the concentration of an oligonucleotide drug product for quality control release as well stability [62]. As more laboratories explore the versatility of CE, method validation paradigms will evolve for the successful application of the various modes of this technique for routine pharmaceutical analysis.

### 9.4.6 Chromatographic Methods

Chromatographic methods are perhaps the most versatile of techniques available to the pharmaceutical analytical chemist with applications in all aspects of drug analysis [36, 37] including identity tests, quantitative measurements for impurities content, limit tests for control of impurities as well as quantitative measurements of the active drug in the drug substance and drug product. Chromatographic tests are also used extensively to monitor the stability of pharmaceutical products by tracking the parent peak along with those of the degradation products. The most common modes of chromatography used for the analysis of biopharmaceuticals are size exclusion chromatography (SEC), reversed phase liquid chromatography (RP-LC), ion exchange chromatography (IEC) and hydrophobic interaction chromatography (HIC). A general description of these techniques is followed by an overall discussion of their validation requirements.

Size exclusion chromatography, a separation based on size, is commonly used for confirming the identity of a biopharmaceutical by co-migration with a reference standard of the protein of interest. It is also used for impurity profiles and quantitative limit tests because of its ability to detect aggregates or fragmentation and, when run under reducing conditions, clipped forms of the parent protein. It is important to note that the separation in size exclusion chromatography is based on the hydrodynamic volume of the protein and not its molecular weight. Thus, estimates of molecular weight information based on generic molecular weight standards are crude at best. The net hydrodynamic volume of a protein is strongly dependent on its secondary structure with relatively minor changes leading to dramatic differences in observed retention times. For this reason, extrapolation of molecular weight information from SEC data should be done with caution. For recombinant proteins that have been thoroughly characterized as to their molecular size, however, SEC can be very helpful in assessing secondary, tertiary and quarternary structures, for example, dimerization versus aggregation.

Reversed phase liquid chromatography may be used directly for the identity and purity analysis of biopharmaceuticals as well as part of peptide mapping studies. Similar to SEC, reversed phase identity tests are conducted by co-migration of the sample with that of a reference standard. In addition to the use of solvent gradients and temperature, selectivity enhancements are often attained by varying the pH of the mobile phase and/or incorporating ion pairing reagents such as trifluoroacetic acid in the aqueous portion of the mobile phase.

Ion exchange chromatography, a separation process based on charge characteristics of a molecule is also used for the identity and impurity profile analysis of biopharmaceuticals. Cation exchange chromatography can detect the occurrence of deamidation or oxidation reactions as well as glycosylation variants and can therefore be used for the stability monitoring of proteins. Hydrophobic interaction chromatography, a multimode technique, refers to separations carried out with hydrophobic sorbents with elution usually involving a decreasing salt concentration. Selectivity is influenced by a number of eluent factors including the type of salt or organic modifier present in the eluent, the addition of surfactants and column temperature. For protein samples, interaction of the non-polar amino acid residues with the hydrophobic stationary phase when combined with the separating power of decreasing salt gradients results in very high resolution separations of deamidation products, methionine sulfoxides, N-terminal variants, glycosylation variants, disulfide isomers and proteolytic clips. Hydrophobic interaction chromatography is also used extensively in identity testing, qualitative and quantitative limit tests and the stability monitoring of pharmaceutical proteins. It is usually the case that more than one form of chromatography is performed on each protein sample and upon satisfactory validation can provide complementary information or confirmatory evidence of identity and/or purity.

An important step in the validation process of a chromatographic method is the definition of appropriate system suitability parameters to ensure successful analysis. They are typically defined as a range of acceptance values for a series of key parameters such as precision, tailing factor, peak resolution, retention time, linearity of detector response and recovery as deemed appropriate for the scope of analysis (See also Chapter 2). Other elements of validation common to all chromatographic techniques are sample preparation protocols, sample injection conditions, mobile phase considerations, column effects and detection systems. These are especially important in the analysis of biopharmaceuticals because the three dimensional structure of proteins is sensitive to minor changes in the environment such as pH, temperature, surface area/nature of the stationary phase, adsorption phenomena, protein concentration and mobile phase composition. Structure changes can dramatically alter the retention characteristics of the protein. Thus, if these parameters are not carefully controlled, they can contribute to poor reproducibility of the retention times as well as observed peak areas.

Sample preparation protocols must be evaluated to confirm good recoveries of the protein by minimizing losses to surface adsorption. Effect of matrix components such as formulation excipients on the recovery and resulting separation of the analyte must be examined. Sample injection parameters such as injection volume, sample matrix and sample concentration may also have an effect on the separation.

The impact of column effects such as temperature effects, mobile phase composition, carryover of analytes, irreversible or non-specific adsorption and column-to-column variability on the overall separation should be examined and documented. Specificity of the method must be established by ensuring that the detector response for the drug is free of interferences from formulation excipients. Detection of process related impurities in the drug sample should also be demonstrated. This is often achieved by recording the chromatograms of authentic samples of potential process related impurities and demonstrating that these peaks are well resolved from the drug. Similarly, the resolution of the drug from its degradation products must be demonstrated by stress testing the drug under forced degradation conditions using acid, base, oxidizing agents and thermal/ photochemical/ enzymatic stress.

Linearity of the detector response for the analyte and other peaks of interest must be established for normal sample concentrations. Validity of using single point calibrations must be demonstrated by identification of the working range at which ratio of detector response to concentration of the sample and the standard are equivalent. It is common practice to assume that the detector response at a concentration is the same for the analyte and the other peaks of interest in the chromatogram (i.e. impurities, degradation products). This assumption may not always be valid and could lead to inaccurate results particularly when used for limit tests and quantitative tests as a measure of drug purity.

Determination of the limit of detection (LOD) and limit of quantitation (LOQ) of impurities must be done in the sample matrix in which the actual analysis will be performed. Also, the LOQ should be within the working concentration range of the analysis. It is important to always report the LOD and the LOQ with their associated S/N values. While the LODs for biopharmaceuticals are commonly reported at S/N = 2 and the LOQs at S/N = 4, these may vary significantly depending on the type of protein and the separation/detection system employed. Recovery studies are often performed by spiking a known concentration of the protein into a placebo formulation. Care should be taken to ensure that the placebo is a true representation of the actual drug product formulation (i.e. the composition and ratio of ingredients are identical to that of the drug product). Recovery studies of impurities and degradation products must also be evaluated preferably by spiking low amounts of authentic samples into actual drug product samples. Peak purities must be established either by collection of the chromatographic peaks followed by structural analysis or by the use of appropriate detection systems such diode array to record the UV-visible spectrum at the beginning, middle and tail of each peak or by on-line LC-MS analysis.

As described in the recent ICH guideline [5], there are several levels of precision. For chromatography, system precision is often determined as part of the development of the system suitability parameters. In addition to system precision, method precision includes reproducibility of sample preparation and sample handling aspects of the analysis. Overall precision of the method encompasses all these factors and is examined for within-day and day-to-day variations and as part of ruggedness or robustness testing of between-operator and between-laboratory variations.

### 9.4.7 Carbohydrate Analysis

Recombinant proteins derived from eukaryotic cell lines are characteristically glycosylated and this glycosylation is known to be cell-line dependent resulting in glycoproteins that may be identical in their peptide sequence but containing significantly different oligosaccharide structures. The degree and type of glycosylation as well as the microheterogeneity may have implications on the activity, potency, stability and safety of a biopharmaceutical, making carbohydrate analysis a requisite part of the quality control testing of biotechnology products derived from eukaryotic cell lines.

Carbohydrate analysis is an inherently challenging task. With current analytical technology, it may not be possible to fully characterize these oligosaccharide structures and the scope of analytical testing is often restricted to demonstrating consistency between different lots of glycoproteins as a realistic alternative.

Carbohydrate analysis to demonstrate product consistency in routine quality control rely on compositional analysis, determination of the glycosylation sites and/or oligosaccharide mapping techniques. Compositional analysis generally involves simple colorimetric tests to quantitate the neutral sugars and sialic acid. The glycosylation sites, whether they are N-linked to asparagine or O-linked to serine or threonine, can be determined by peptide mapping techniques [23] and by mass spectrometry [25]. However, microheterogeneity of the oligosaccharide chains can greatly influence the LC separation in peptide mapping resulting in broad peaks that may not be amenable to meaningful interpretation.

Another approach to characterize these oligosaccharide side chains is to cleave them from the protein polypeptide chain by chemical or enzymatic means, followed by analysis by anion-exchange chromatography under high pH conditions. Because of lack of a sensitive chromophore, electrochemical detection is generally used in the pulsed amperometric mode to generate a carbohydrate map that serves as a fingerprint not unlike a peptide map. This characteristic fingerprint is then compared with that of the reference standard to establish product consistency. A necessary element of validation of such mapping techniques is structural characterization of the peaks in the carbohydrate map as well as demonstration of peak purities. Recent advances in LC-MS [38] are making such structural characterization possible but these novel techniques are still in the development stage and are primarily used as part of the initial characterization of a reference standard or other specialized applications.

### 9.4.8 Protein Content

A quantitative protein content assay is often one of the first methods to be validated due to its importance in the early development of isolation, separation and purification processes of a pharmaceutical protein. This measurement is used for a variety of applications during the product development stage including assessment of specific activity, determination of protein recovery from various steps of the manufacturing process, formulation of preclinical and clinical supplies and the establishment of a reference standard. Measurement of protein content is an important element of the purity analysis of protein pharmaceuticals as *every* weight-for-weight measurement, whether impurities or related compounds, is based on the protein content of a given sample. Also, the potency of each lot of drug product is expressed as the activity per unit mass in which the unit mass represents actual protein content. There are many approaches that may be used to measure protein content and it is often necessary to use multiple techniques to confirm accuracy.

#### a. Absolute Assays

One of the simplest methods for determining protein content involves measurement of the ultraviolet absorbance of the sample at the absorption wavelength maximum of the protein. For proteins containing aromatic amino acid residues such as tyrosine, tryptophan or phenylalanine, a detection wavelength of 280 nm is generally used. The observed absorbance may be converted to protein concentration by application of Beer's Law. Because of the influence of a number of common experimental conditions such as ionic strength, pH, buffer salts, solvents and other excipients on absorbance, the molar absorptivity of the reference standard must be determined in a solvent medium which is identical to that of the sample. Furthermore, this type of measurement is only valid for very pure samples, since the presence of any related protein impurities that have appreciable absorbance at the detection wavelength would be recorded as the protein of interest leading to a higher than true protein content. Thus a chromatographic or electrophoretic experiment to assure that the sample is free of these impurities (or quantitate and adjust the UV data) is necessary to assure accuracy.

Another factor that could affect the accuracy of UV absorbance measurements is the possibility of protein aggregation, which could result in light scattering leading in turn to erroneously high absorbance readings. This may be readily overcome by dilution of the sample or addition of reagents known to minimize such aggregation. Of course, if such reagents are used, their impact on molar absorptivity must be established as part of the validation process. Once the above parameters have been thoroughly evaluated and the method validated, UV-visible spectrometry remains the simplest and most straightforward measure of protein content for a variety of applications during and post manufacture of pharmaceutical proteins. It is also an absolute measurement without the need for comparative analysis against a pre-established reference standard.

Another absolute assay of protein content is the Kjeldahl assay, which involves the complete destruction of the protein by digestion with sulfuric acid. The nitrogen from the protein is trapped as ammonium sulfate and the ammonia is steam distilled from the hydrolysate after basification with NaOH. The concentration of ammonia is then determined by either titration or calorimetric means or an ammonia selective electrode from which the amount of protein in the original sample may be calculated. While this method is accurate and precise it is labor intensive and requires relatively large quantities (> 1 mg) of sample. To assure accuracy of the protein content determination, validation of this method requires optimizing the experimental conditions so that digestion reaction proceeds to completion. For proteins that are difficult to digest, addition of inorganic salts such as potassium sulfate or the use of alternate oxidizing agents may prove helpful. The presence of amino sugars, nitrogen containing excipients and other matrix components can also have an impact on the quantitation of total protein. This is particularly a problem with glycoproteins containing oligosaccharide side chains that are digested by reaction with sulfuric acid to yield a higher than expected value for the protein content (i.e. positive bias). For these reasons, the results of this method should be cross validated against other protein content assays to assure accuracy.

A classical biochemical technique is gravimetric or dry weight analysis in which the difference in the dry weight of a sample is recorded before and after desalting. This difference in weight is attributed to the amount of protein in the sample. While this is also considered an absolute method and is generally of relatively high precision, its accuracy is highly dependent upon the effectiveness of the desalting and drying procedures. This is especially a problem for hygroscopic proteins that have bound moisture that is difficult to remove completely. Therefore the resulting protein content values are highly suspect unless crossvalidated against results from other methods.

Quantitative amino acid analysis is one of the most accurate, precise and reliable methods for determining the protein content of a biopharmaceutical [39-41]. To determine protein concentration, the data from the amino acid analysis experiment are plotted as the observed nmoles of each amino acid residue versus the theoretical nmoles of each residue per nmole of protein. The slope of this plot yields protein concentration information upon correction for injection volumes and dilution factors [42]. Validation considerations for protein content determination by this method are essentially the same as those discussed at length for amino acid analysis.

### b. Relative Assays

One of the most common assays in this category is the Lowry assay which is based on the Biuret reaction of proteins with a cupric solution to form a protein-Cu(II) complex under basic conditions. Oxidation of the aromatic amino acids tyrosine, tryptophan and phenyalanine in the protein reduces the Cu(II) to Cu(I) which, upon reaction with Folin-Ciocalteau reagent, gives rise to an intense blue color which is monitored spectrophotometrically between 540 and 560 nm [43]. While this is a relatively simple method, its susceptibility to interferences from matrix components such as buffer salts, sugars and alcohols may result in inaccurate results. Accuracy may be improved significantly, however, by identification and elimination of these interferences or by the use of an identical matrix for the standards as that of the sample to compensate for these effects. Also, because the product of the Lowry reaction is photosensitive, the length of time between reaction and reading of the UV absorbance may have a significant impact on the accuracy as well as the reproducibility of this measurement. Accuracy may also be enhanced by using reference standards of the protein of interest to generate the standard curves instead of the commonly used BSA. The method can be validated readily to yield relatively high accuracy with LODs typically in the range of 5 - 10 μg/mL. However, one of the major drawbacks of the Lowry assay is the poor precision afforded by the method's inherent sensitivity to minor changes in the experimental protocol.

A somewhat more rugged method is the bicinchoninic acid (BCA) assay in which a protein-Cu(II) complex is reduced to Cu(I) and the resultant complex monitored spectrophotometrically at 560 nm [44]. This method is considerably easier to validate as it is a single step procedure and the reagents used are much more stable to general handling. Also, this assay has a much higher sensitivity and precision than the Lowry assay due to fewer interferences from common sample matrix components. This assay is amenable to adaptation to a microtiter plate format and the use of automated plate readers yielding reproducibilities in the order of 2 - 3 % RSD within day. The limits of detection are in the range of 5 to 1000 μg/mL.

Another colorimetric assay in common use is the Bradford Assay in which the protein forms a complex with the dye Coomassie Brilliant Blue G-250 under acidic conditions resulting in a shift in the absorption maximum of the dye from 465 nm to 595 nm [45]. The reaction is complete in approximately 2 minutes and the product is stable for about an hour. The concentration of the protein is determined by comparison of the absorbance of the protein reference standard with that of the sample. In addition to its simplicity and quick turnaround, this assay is also highly sensitive with detection limits to <1 μg/mL. Because it is relatively insensitive to sample matrix interferences, its precision is generally good. However, validation of this assay poses some challenges. The reaction products are susceptible to degradation, so the length of time between the Coomassie Blue reaction and reading the absorbance must be carefully controlled. As the standard curves tend to be non-linear it is essential to work within the linear range of the assay, which varies from protein to protein. In addition, the dye is known to bind to the surface of cuvettes, the extent of which may be dependent upon the concentrations of the dye, the protein and other variables.

Fluorescence-based assays have become quite popular due to their inherent sensitivity, which can improve the limits of detection by approximately three orders of magnitude over UV detection. By far the most common fluorogenic reagent is *o*-phthalaldehyde (OPA) with an excitation wavelength of 340 nm and an emission maximum at 450 nm. OPA reacts rapidly with primary amines in the presence of a thiol to form a fluorescent product that is reasonably stable in aqueous buffers and organic solvents. Because of its solubility in organic solvents, it is particularly useful for determining the protein content of hydrophobic proteins. Validation concerns of an OPA based protein content assay include the need to avoid tris and amino acid buffers which react with OPA/thiol. This may effectively preclude the use of this method for many samples. The reagent reacts primarily with lysine residues and may require hydrolysis of the protein to yield accurate results. Another fluorescence agent of similar assay characteristics is fluorescamine requiring much of the same approaches as the OPA/thiol reagent system to yield good accuracy and precision.

### 9.4.9 Biological Activity/Potency Assays

Biological activity assays are critical to pharmaceuticals derived from recombinant protein technology, highlighting the inadequacy of current analytical technologies to assure pharmacological activity on the merits of physical chemical characterization of a protein (i.e. it is possible that a particular lot of the drug product will pass all physical chemical criteria and still not produce the desired pharmacological effect *in vivo*). This is due to the inherent complexity of a macromolecule whose activity is dependent upon not just the primary sequence but also its secondary, tertiary and quaternary structures under physiological conditions. Hence, there is a necessity for establishing a primary reference standard against which the biological activity of each lot of a biopharmaceutical may be assayed.

In general, a primary reference standard is the purest lot of drug substance available in sufficient quantity to perform a thorough physical, chemical and biological characterization. For conventional pharmaceuticals manufactured by chemical synthesis the reference standard is established by the manufacturer of the pharmaceutical. In the case of a biopharmaceutical, the primary reference standard may be established in collaboration with the FDA and other regulatory agencies and will become the internationally recognized measure of potency for that drug. Often the reference standard is not of recombinant origin but an isolated natural product with the desired pharmacological activity. Samples of these standards with a well defined biological activity (in international units/vials) may be obtained from a variety of sources such as the World Health Organization (WHO), United States Pharmacopeia (USP), National

Institute for Biological Standards and Control (NIBSC) and the National Institutes of Health (NIH). In practice, a working reference standard is often established by assay against the primary reference standard for use in routine analysis. All reference standards used for potency assessment must be traceable to a primary reference standard from one of the above mentioned agencies.

According to the Code of Federal Regulations, "*Tests for potency shall consist of either in vitro or in vivo tests, or both, which have been specifically designed for each product so as to indicate its potency*". There are currently a number of approaches for the potency assessment of a biopharmaceutical. They include "living model assays" using whole animal or cell culture bioassays, specific binding assays consisting of immunoassays or receptor binding assays and finally, biochemical or physical-chemical assays such as LC.

**a. Whole animal bioassays**

In this approach, live animals that are capable of expressing the desired biological effect in man are employed to ascertain that the biopharmaceutical sample of interest has statistically significant relative potency when compared with that of the reference standard. Whole animal bioassays are generally used only because an alternative *in vitro* or biochemical assay has not been demonstrated to be predictive of the activity of the drug of interest. Live animal assays are difficult to validate requiring large numbers of animals per dose/treatment group due to poor animal-to-animal reproducibility. Precision in the range of 30 - 100 % RSD are not uncommon. In addition, the validation efforts required are expensive and require extraordinary lengths of time as each bioassay requires several days or several weeks to detect a measurable biological effect after dosing with the drug. Although validation of such assays is achievable as evidenced by the whole animal bioassays that have been established for biotechnologically derived products such as human growth hormone [46], and human insulin [47], it is highly desirable to develop alternative systems for establishing the potency of a pharmaceutical protein.

**b. Cell Culture Based Bioassays**

Cell culture based bioassays rely on the binding ability of the drug to its receptor on the cell resulting in a measurable biological effect. A major assumption of this assay is that a measurable response at the cellular level can be related to therapeutic activity *in vivo*. In order to meet the criterion for accuracy, this assumption should be validated by comparison of the *in vitro* results with *in vivo* activity in suitable whole animal models. Cell culture based assays can also be highly selective with the potential to detect structural variants in the sample. To confirm this, authentic samples of the structural variants of interest should be evaluated in this assay and demonstrated to yield little or no (or possibly enhanced) activity under the conditions of the assay. The precision of cell culture based bioassays is typically in the range of 20 - 30 % RSD, which, although considered poor for many conventional analytical techniques, is a significant improvement over whole animal assays. Imprecision arises primarily from inherent variability in the living cells used but, due to this assay's ready amenability to automation, relatively short analysis times and moderate costs, appropriate number of replicate tests may be performed to yield precision values in an acceptable range.

Living cells are also sensitive to sample handling and the types of solvents, buffers salts, ionic strength, drug concentration, temperature and pH can have a significant impact on the accuracy and precision of these assays. In addition, one million fold dilutions of the drug are not uncommon making laboratory techniques a key factor in assuring good reproducibility. Upon suitable validation to yield consistently accurate potency information, cell culture based bioassays provide a much more desirable and viable alternative to the whole animal bioassays discussed earlier. Successful cell culture assays have been developed for erythropoietin [48], Interleuken-2 [49], α-interferon [50], human interferon β, [51] G-CSF [52], and GM-CSF [53, 54].

### c. Immunoassays and Receptor Binding Assays

Immunoassays and receptor binding assays are similar to cell culture based bioassays in that they rely on antigen-antibody or receptor-ligand binding as a measure of potency. These *in vitro* assays are yet another step removed from true activity measurements as the relevance of binding activity to therapeutic activity *in vivo* may be difficult to establish. In addition, the assays often cannot discriminate between biologically active versus biologically inactive material due to significant cross reactivity with related agents. However, its simplicity, minimal cost and amenability to automation renders this approach attractive for potency determination. Immunoassays can provide excellent sensitivity and precision, particularly if automated, but their lack of specificity and its impact on accuracy must be clearly documented and validated prior to its use as a sole measure of potency. Receptor binding assays have been shown to be good indicators of human Interleukin-1α [55], Interleukin-2 [56], and GM-CSF [57] activity. An antigen-antibody immunoassay is currently being used to predict the activity of murine monoclonal antibody.

### d. Biochemical Methods

Biochemical assays are based on the chemical action of a biopharmaceutical which mimics the biological function. A good example of such an assay is that for tissue plasminogen activator (TPA) in which the biological function of TPA (lysing of blood clots) is mimicked by the addition of TPA and plasminogen to a synthetic fibrin clot [58, 59]. The time required for lysis of the synthetic clot by plasmin produced by TPA's activation of plasminogen is measured by spectrophotometry. Because the measured parameter is a chemical signal, the precision of this type of assay is quite good. It is also amenable to automation, relatively inexpensive and simple to execute when compared with bioassays. However, the most important criteria for the validity of such an assay is its accuracy in mimicking the biological function and its sensitivity to the presence of protein variants in the sample. This would require comparison of the results of this assay against a valid whole animal or cell culture based bioassay as well as measure its response to authentic structural variants of the target protein.

### e. Physical-Chemical Assays

These assays rely on the ability of physical chemical methods to predict *in vivo* activity. This is extremely difficult to accomplish due to the inherent complexity of a macromolecule consisting of significant secondary, tertiary and quaternary structures. It is generally the case that biological function is much more sensitive to minor changes to such secondary structures than a typical physical chemical technique. Therefore, validation of such a physical chemical technique to be a true indicator of potency requires extensive validation against reliable whole animal and/or cell culture based bioassays. Although difficult to develop, upon demonstrable relevance and indication of biological activity, they are highly desirable due to their high precision, potential for automation, cost and rapid analysis times. An example of such an assay is the LC assay of human insulin which has been extensively validated to be a good predictor of biological activity [60, 61].

## 9.5. Emerging Technologies

In addition to recombinant proteins and monoclonal antibodies, recent advances have led to a plethora of new technologies such as antisense oligonucleotides, gene therapy and therapeutic vaccines. The mechanism of action of the drug candidates resulting from these technologies pose unique challenges in drug delivery often requiring novel formulations optimized for penetration of the cell wall and resistance to intracellular components. While many of these challenges are being addressed *via* conventional drug delivery systems by incorporation of the active drug into structures such as liposomes, a new arena of novel drug delivery systems, such as virus mediated cell penetration agents, have emerged to address some of the unique delivery requirements of these novel biomolecules. As these technologies mature to generate viable candidates for drug development, there will arise a need to characterize these novel therapeutic agents in complicated drug delivery systems. In addition, the potency of many of these drugs are, by design, significantly higher than those of conventional drugs, posing new demands for highly sensitive analytical methods for bioanalytical applications. This has resulted in a fundamental change in the nature of the analytical development group from primarily chemists to a team of molecular biologists, pharmacologists, microbiologists, virologists and the like.

A good example of an emerging technology is the area of oligonucleotide or antisense therapeutics. Many of the techniques for characterizing these molecules are modifications of already existing methods for peptide and protein analysis. But chemical modifications either in the phosphodiester backbone or in the nucleobases towards *in vivo* stabilization of oligonucleotides renders many of the common analytical techniques used for nucleotide analysis inadequate. Thus new paradigms are being developed for the analysis of these novel antisense oligonucleotide therapeutic agents incorporating the use of chromatography and capillary gel electrophoresis for impurity profiles, [62] modified Maxam-Gilbert [63] or mass spectrometry [64] for sequencing and modified protocols for base composition analysis [65]. It has also been noted that these oligonucleotides are inhibitory in the conventional gel clot assay for endotoxin necessitating alternate approaches to assure microbiological quality [66].

While the underlying principles of validation such as accuracy, precision, reliability and ruggedness are common to all analytical technologies, the uniqueness of each new method will dictate the validation criteria warranting use for routine analysis. As the more promising candidates move into active clinical development and on to product registration, there are unprecedented opportunities for the pharmaceutical analytical chemist in the development of new analytical technologies to meet the needs in all facets of development from identification of the drug candidate for development through product registration and post market support.

## 9.6 Common sense approach to regulatory concerns: Partnership with the FDA

The past decade has seen an explosion of new technologies, a significant number of them reaching the stage of new product registrations in the next few years. Many of the new technologies have elements of both drugs and biologicals clouding expectations of regulatory requirements for drug approval. Regardless of the nature of the technology leading to the therapeutic entity in question, the primary expectation of any regulatory body is statistically significant proof of safety and efficacy. For therapeutic entities derived from these new technologies, the specifics of the regulatory requirements are best delineated in a partnership arrangement between the pharmaceutical industry and the regulatory agency. The effectiveness of such a collaborative approach has been demonstrated in the development of "Points to Consider" documents towards registration of drug products from recombinant DNA technology and monoclonal antibodies. For future products of biotechnology this approach will allow development of realistic guidelines for assuring the quality of the drug product by tailoring the regulatory requirements around the strengths and limitations of current analytical technology.

## References


1. Y.-Y.H. Chiu, in *Biotechnologically Derived Medical Agents. The Scientific Basis of their Regulation.* (J. L. Gueriguian, V. Fattorusso and D. Poggiolini, Eds.), pp. 1-22, Raven Press, (1988)

2. E. Esber, in *Biotechnologically Derived Medical Agents. The Scientific Basis of their Regulation.* (J. L. Gueriguian, V. Fattorusso and D. Poggiolini, Eds.), pp. 163-67, Raven, (1988)

3. Guideline for Submitting Documentation for the Stability of Human Drugs and Biologics, Center for Drugs and Biologics, FDA (1987)

4. Guideline for Stability Testing of New Drug Substances and Products, International Council on Harmonization, FDA (1994)

5. Guideline on Validation of Analytical Procedures for Pharmaceuticals; Availability, International Council on Harmonization, FDA (1994)

6. Points to Consider in the Production and Testing of New Drugs and Biologicals Produced by Recombinant DNA Technology, Office of Biologics Research and Review, FDA (1985)

7. Points to Consider in the Manufacture of Monoclonal Antibody Products for Human Use, Office of Biologics Research and Review, FDA (1987)

8. Points to Consider in the Characterization of Cell Lines Used to Produce Biologicals, Office of Biologics Research and Review, FDA (1987)

9. V.R. Anicetti, B.A. Keyt and W.S. Hancock, *Trends Biotech.*, **7**, 342-9 (1989)

10. L.J. Schiff, W.A. Moore, J. Brown and M.H. Wisher, *Biopharm.*, **5**, 36-9 (1992)

11. E. Joner and G.D. Christiansen, *Biopharm.*, **1**, 50-5 (1988)

12. J. Briggs and P. Panfili, *Anal. Chem.*, **63**, 850-9 (1991)

13. V.T. Kung, *Anal. Biochem.*, **187**, 220-7 (1990)

14. *Code of Federal Regulations.*, **21,** Section 610.12, (1995)

15. *United States Pharmacopeia National Formulary, 23rd edn.* pp. 1718-9, USP, Rockville (1995)

16. *United States Pharmacopeia National Formulary, 23rd edn.* pp. 1696-7, USP, Rockville (1995)

17. Guideline on the Validation of the Limulus Amebocyte Lysate Test as an End-Product Endotoxin Test for Human and Animal Parenteral Drugs. Biological Products and Medical Devices (1987)

18. *United States Pharmacopeia National Formulary, 23rd edn.* pp. 1686-90, USP, Rockville (1995)

19. A.L. Burlingame, T.A. Baillie and D.H. Russell, *Anal. Chem.*, **64**, 467R-502R (1992)

20. J.B. Smith, G. Thevenon-Emeric, D.L. Smith and B. Green, *Anal. Biochem.*, **193**, 118-24 (1991)

21. R.L. Garnick, *Dev. Biol. Stds.*, **76**, 117-30 (1992)

22. J. Dougherty Jr., L.M. Snyder, R.L. Sinclair and R.H. Robins, *Anal. Biochem.*, **190**, 7-20 (1990)

23. J.J. L'Italien, *J. Chromatogr.*, **359**, 213-20 (1986)

24. J. Frenz, J. Bourell and W.S. Hancock, *J. Chromatogr.*, **512**, 299-314 (1990)

25. V. Ling, A.W. Guzzetta, E. Canova-Davis, J.T. Stults, W.S. Hancock, T.R. Covey and B.I. Sushan, *Anal. Chem.*, **63**, 2909-15 (1991)

26. D.A. Lewis, A.W. Guzzetta and W.S. Hancock, *Anal. Chem.*, **66**, 585-95 (1994)

27. A.S. Bhown, J.E. Mole, A. Weissinger and J.C. Bennett, *J. Chromatogr.*, **148**, 532-35 (1978).

28. C.L. Zimmerman, E. Apella and J.J. Pisano, *Anal. Biochem.*, **77**, 569-73 (1977)

29. R.T. Swank and K.D. Munkres, *Anal Biochem*, **39**, 462-77 (1971)

30. B.D. Hames and D. Rickwood, in *Gel Electrophoresis of Proteins: A Practical Approach*. (B.D. Hames and D. Rickwood, Eds.), pp. 1-143, IRL, Washington D.C. (1983)

31. C.R. Merril, R.C. Switzer and M.L. VanKeuren, *Proc. Natl. Acad. Sci. USA*, **76**, 4335-9 (1979)

32. C.M. Manning, K. Patel and R.T. Borchardt, *Pharm. Res.*, **6**, 518-23 (1989)

33. F. Wold, *Ann. Rev. Biochem.*, **50**, 783-814 (1981)

34. P.D. Grossman, H.H. Lauer, S.E. Moring, D.E. Mead, M.F. Oldham, J.H. Nickel, J.R.P. Goudberg, A. Krever, D.H. Ransom and J.C. Colburn, *Amer. Bio. Lab.*, Feb., 35-43 (1990)

35. A.D. Tran, S. Park, P.J. Lisi, O.T. Huynh, R.R. Ryall and P.A. Lane, *J. Chromatogr.*, **542**, 459-71 (1991)

36. J. Frenz, in *Chromatography in Biotechnology*. (C. Horvath and L. S. Ettre, Eds.) **529**, pp. 1-12, ACS, Washington, D.C. (1993)

37. J. Frenz, W.S. Hancock, W.J. Henzel and C. Horvath, in *HPLC of Biological Macromolecules: methods and applications*. (K. Gooding and F. Regnier, Eds.) **51**, pp. 145-77, Dekker, NY (1990)

38. A.W. Guzzetta, L.J. Basa, W.S. Hancock, B.A. Keyt and W.F. Bennett, *Anal. Chem.*, **65**, 2953-62 (1993)

39. R.L. Lundblad and C.M. Noyes, in *Chemical Reagents for Protein Modifications.*, CRC Press, Boca Raton (1984)

40. S. Blackburn, in *Amino Acid Determination: Methods and Techniques*, pp. 11-15, Dekker, New York (1978)

41. C.W. Gehrke, L.L. Wall, J.S. Absheer, F.E. Kaiser and R.W. Zumwalt, *J. Assoc. Off. Anal. Chem.*, **68**, 811-21 (1985)

42. R.R. Granberg, *Amer. Biotechnol. Lab.*, June, 58-69 (1984)

43. O.H. Lowry, N.J. Rosebrough, A.L. Farr and R.J. Randall, *J. Biol. Chem.*, **193**, 265-75 (1951)

44. P.K. Smith, R.I. Krohn, G.T. Hermanson, A.K. Mallia, F.H. Gartner, M.D. Provenzano, E.K. Fujimoto, N.M. Goeke, B.J. Olson and D.C. Klenk, *Anal. Biochem.*, **150**, 76-85 (1985)

45. M.M. Bradford, *Anal. Biochem.*, **72**, 248-54 (1976)

46. D.R. Bangham, R.E.G. Das and D. Schulster, in *Hormone Drugs: Proceedings of the FDA-USP Workshop on Drug and Reference Standards for Insulins, Somatotropins, and Thyroid-axis Hormones*. pp. 301-12, United States Pharmacopeial Convention, Inc., Bethesda (1982)

47. A.F. Bristow, R.E.G. Das and D.R. Bangham, *J. Biol. Standard.*, **16**, 165-78 (1988)

48. N. Imai, M. Higuchi, A. Kawamura, K. Tomonoh, M. Oh-Eda, M. Fujiwara, Y. Shimonaka and N. Ochi, *Eur. J. Biochem.*, **194**, 457-62 (1990)

49. T. Yamada, A. Fujishima, K. Kawahara, K. Kato and O. Nishimura, *Arch. Biochem. Biophys.*, **257**, 194-9 (1987)

50. T. Arakawa, Y.R. Hsu, M.A. Narachi, M.F. Rohde and P. Hennigan, *Biopolymers*, **29**, 1065-8 (1990)

51. J. Geigert, D.L. Ziegler, B.M. Panschar, A.A. Creasey and C.R. Vitt, *J. Interferon Res.*, **7**, 203-11 (1987)

52. M. Oh-eda, M. Hasegawa, K. Hattori, H. Kuboniwa, T. Kojima, T. Orita, K. Tomonou, T. Yamazaki and N. Ochi, *J. Biol. Chem.*, **265**, 11432-5 (1990)

53. P. Moonen, J.J. Mermod, J.F. Ernst, M. Hirschi and J.F. DeLamarter, *Proc. Natl. Acad. Sci. U S A*, **84**, 4428-31 (1987)

54. R.E. Donahue, E.A. Wang, R.J. Kaufman, L. Foutch, A.C. Leary, J.S. Witek-Giannetti, M. Metzger, R.M. Hewick, D.R. Steinbrink

and G. Shaw, *Cold Springs Harbor Symp. Quant. Biol.*, **51**, 685-92 (1986)

55. F. Riske, R. Chizzonite, P. Nunes and A.S. Stern, *Anal. Biochem.*, **185**, 206-12 (1990)

56. K. Kitamura, K. Matsuda, M. Ide, T. Tokunaga and M. Honda, *J. Immunol. Methods*, **121**, 281-8 (1989)

57. C.B. Brown, C.E. Pihl and K. Kaushansky, *European J. Biochem.*, **225**, 873-80 (1994)

58. M. Christodoulides and D.W. Boucher, *Biologicals*, **18**, 103-11 (1990)

59. R.L. Garnick, *J. Pharm. Biomed. Anal.*, **7**, 255-66 (1989)

60. *United States Pharmaceopeia National Formulary, 23rd edn.*, pp. 809-811, United States Pharmaceopeial Convention, Inc., Rockville (1995)

61. H.W. Smith, L.M. Atkins, D.A. Binkley, W.G. Richardson and D.J. Miner, *J. Liq. Chrom.*, **8**, 419-39 (1985)

62. G.S. Srivatsa, M. Batt, J. Schuette, R.H. Carlson, J. Fitchett, C. Lee and D. L. Cole, *J. Chromatogr.A.*, **680**, 469-77 (1994)

63. T. K. Wyrzykiewicz and D.L. Cole, *Nucl. Acids Res.*, **22**, 2667-9 (1994)

64. J.M. Schuette, U. Pieles, S.D. Maleknia, G.S. Srivatsa, D.L. Cole, H.E. Mozer and N.B. Afeyan, *J. Pharm. Biomed. Anal.*, **13**, 1195-1203 (1995)

65. J.M. Schuette, D.L. Cole and G.S. Srivatsa, *J. Pharm. Biomed. Anal.*, **12**, 1345-53 (1994)

66. M.A. Feldman, J.M. Parsons, J.A. Lee and G.S. Srivatsa, Abstract No. M-P/D5, *Sixth Int. Symp. Pharm. Biomed. Anal.*, St. Louis, (1995)

# Chapter 10

# Biological Samples

*Krzysztof A. Selinger*

## 10.1 Introduction

The subject of bioanalytical method validation has been debated extensively for the last several years and a number of papers describing various approaches to validation have been published [1-10]. The need to develop and accept a uniform approach to validation has been generally recognized as important in the pharmaceutical industry. This has resulted in two meetings and two widely quoted publications: 1) the report of the Conference on "Analytical Methods Validation: Bioavailability, Bioequivalence and Pharmacokinetic Studies" held in Washington in December 1990 [11], and 2) Drug Information Association's Workshop on International Consensus Statement on Bioavailability/Bioequivalence Testing Requirements and Standards held in Barcelona in March of 1991 [12]. Numerous workshops and follow-up meetings ensued. The two reports mentioned above will be referred to as the Conference Report and the DIA Consensus; collectively they will be referred to as consensus reports.

These two documents are very similar and should be used as a base for any bioanalytical validation; however, it should be remembered that they represent a minimum requirement in the industry. It should be also understood that some of the consensus recommendations do not necessarily enjoy the full support of the pharmaceutical community [13] and other approaches will naturally evolve with the progress of analytical methods and techniques. In particular the advent of robust and dependable LC/MS equipment may especially change the face of bioanalysis in the near future.

The nature of the consensus reports is rather general and they do not answer many detailed questions. The intention of this chapter is to elaborate on the consensus reports, to evaluate critically some aspects of validation and to provide detailed and pragmatical information on how to plan and carry out a validation and a bioanalytical project, which would meet current industrial standards. Most of the comments are related to chemical (mainly chromatographic) methods, while issues related to immunoassays will be addressed only in sec. 10.5. The scope will be limited to bioavailability, bioequivalence, and pharmacokinetic studies, plus some aspects of toxicology and metabolism.

The term "Bioanalytical method validation" can be understood in a narrow and broad sense; this chapter will cover the broad sense which includes:

- the proper validation exercise
- application of the validated method to routine drug analysis
- proper execution of a bioanalytical project
- control of a method during its execution

## 10.2 Logistics of Method Validation

One of the critical issues that has not been addressed by the consensus reports is when validation is necessary. In the current highly cost-conscious environment, the balance of costs and benefits is at issue. New drug candidates, frequently numerous analogues of similar structure, are screened and tested and it would be wasteful to undertake a full validation for each of them. Hence, some initial pilot toxicology projects can be done using a method that has not yet been fully validated. Usually at the discovery stage of drug development doses administered to laboratory animals are rather high which enables simple approaches, such as protein precipitation as a rough sample pre-treatment and a robust reversed-phase liquid chromatography with a UV or a fluorescence detector. This is not to say that such a method is without scientific merit and absolves the analyst either from understanding the chemistry of the compounds investigated or from testing at least some rudimentary parameters such as stability (is the compound stable for 1 day or 1 week that is needed to complete this pilot project?) or extraction efficiency (is recovery at least 20-30%?). At this stage a one-day validation procedure consisting of a single or duplicate calibration curves and a set of quality controls is probably sufficient.

Projects leaving the discovery stage and entering the development stage require fully validated analytical methods to support formal toxicology studies. Although the ADME studies (absorption, distribution, metabolism, and elimination) of the parent drug do not require a formal validated method, such a method usually exists at this rather late stage of drug development. All human studies submitted to regulatory authorities require a fully validated method.

Every laboratory needs to develop and adopt a validation protocol that describes the specific elements of validation and acceptance criteria in a particular laboratory. This protocol usually takes the form of a Standard Operation Procedure (SOP). The validation exercise must be auditable and should be concluded by a validation report. A validation report may contain a detailed description of the analytical method, or the method may be described in a separate document such as a SOP or a method sheet.

## 10.3 Development and Initial Validation of a Chemical method

Validation follows methods development, transfer, or modification, and is followed by a method application, i.e. a bioanalytical project. While validation logically follows method development, the method application does not necessarily directly follow validation. A continuity between these two elements is needed. This means that if a method is not used on a regular basis it may have to be revalidated albeit in an abbreviated fashion. The longest period of time elapsed between the use of a bioanalytical method without revalidation should be specified in the respective SOP, and should not extend over a period of few months. The reason for this precaution is that over that period of time changes may occur in a chromatographic system or its component, reagents, analytical personnel or their skills.

A bioanalytical method cannot be developed and validated without considering the ultimate objectives of the intended study. This means that during method development one must keep in mind basic requirements and numerous details which may have bearing on the project: range of concentrations needed and a potential limit of quantitation (LOQ), matrix to be used, anticoagulants in the case of blood or plasma (e.g. EDTA - what concentration ?, heparin - which salt?), volume of blood needed or available per assay (pediatric studies and studies using small animals provide small sample volumes, usually <0.5 mL), stability, and safety considerations. The stability considerations are of utmost importance. For new chemical entities (NCEs) introduced for the first time into animals the stability data in biological matrices are frequently not available and neither is the method. One must immediately develop a skeleton of an assay and employ it to evaluate potential stability problems. Having potential stability or instability issues at least partially resolved one must then make sure that appropriate sample collection and storage procedures are used in clinics or animal rooms.

After a period of trial and tests a bioanalytical method is ready to be validated. The analyst must be sure that experimentation is completed and no more changes will be introduced. Introduction of new changes and modifications will stop the validation process and may require starting of the validation procedure from the beginning.

The consensus reports specify that the essential parameters which need to be defined to ensure the acceptability of a bioanalytical methods are stability, precision, accuracy, sensitivity, specificity, response function and reproducibility; the DIA Consensus also adds recovery. An example of validation scheme is as follows:

- four precision and accuracy batches, which are also used to determine response function, specificity, and LOQ
- stability testing
- recovery (i.e. extraction efficiency) study

### 10.3.1 Precision and Accuracy

The definitions of precision and accuracy are presented in Chapter 2. The goal of this section is to describe all the steps necessary to achieve precision and accuracy appropriate for trace analysis in biofluids. The majority of analytical measurements are relative in nature, which means that results are obtained by comparing sample responses with those of authentic standards. Hence, the quest for accuracy begins with a reference standard [9].

The best reference standard for well-established and easily available drugs should be USP material with enclosed certificate of analysis (COA). Standards from reputable commercial suppliers with COA are also acceptable. New chemical entities are available only from their originators, who should also provide a COA. Nevertheless, it is responsibility of a user to have a COA. Ultimately, if no COA is available, the user must prepare a COA by performing a number of tests to confirm the identity and the purity of the standard; these tests may be spectral, thermal and elemental analyses, LC area summation, water and ash contents, and residual solvents. One must often limit testing to a necessary minimum as material may be in short supply. Use of drug formulations as a source of analytical standards should be strongly discouraged. If a reference standard is difficult to obtain or purchase and a drug formulation seems to be the only source of the material then such a secondary standard should be also characterized analytically by other techniques.

The next step is to ensure the accuracy of the calculations. Many substances exist in forms of salts and/or hydrates at different degrees of purity. Measurements in biological matrices should provide a result expressed in terms of a free substance. The calculations should be verified by a second scientist.

Weighing of the reference standard should be carried out on a properly maintained and currently calibrated analytical balance, with sensitivity of at least 0.00001 gram. In practical terms an amount between 1 and 10 mg will be weighed. Considering the fact that a potential weighing error could bias all subsequent results and in order to avoid an influence of some peculiarities in the analytical technique, two analysts should prepare total of 3 weighings and 3 discrete stock solutions. After dilutions are made by the primary and secondary analysts, these stock solutions should be compared using an appropriate analytical technique such as LC or GC. The acceptance criteria for stocks/spiking solutions should be specified a priori in the unit's SOP; commonly stock solutions giving a response factor within 3-5% from one another are acceptable.

The equation to calculate these differences is as follows:

$$\frac{(x_2 - x_1)}{\overline{x}} . 100\% \qquad (10.1)$$

where $x_1$ is the peak height (area) of the first solution, $x_2$ is the peak height (area) of the second solution, and the $\overline{x}$ is the mean of the two.

Should the differences be greater than the acceptable limits, records of preparation should be reviewed. New stock solutions may be prepared and an outlier ultimately identified and eliminated using an appropriate statistical approach.

Having made sure that the spiking solutions accurately reflect the analyte concentration one stock solution could be chosen to spike both calibration standard and quality controls (QCs). While it is common in the industry to use separate spiking solutions for the calibration standards and for the quality controls, it is opinion of this author, that having verified the correctness of stock solutions prepared as described above, it is no longer necessary to use 2 sets of stock solutions. To the contrary, it may introduce an additional bias.

Spiking of calibration standards and quality controls should involve both a primary and a secondary analyst. The primary analyst should spike the calibration curve, and the secondary should spike the quality control samples, or a combination of thereof. Again, the secondary analyst is needed to make sure that a slightly different analytical technique leads to the same result. This process appears trivial, yet spiking into biological matrices provides ample opportunity for technical diversity and difficulty. Examples of this are described below.

First, one may choose to use either volumetric flasks where volumes are restricted to certain values or deliver volumes using a pipet as biological matrices tend to be precious and should not be wasted; hence the smallest necessary volume should be used. Second, blood, plasma, and serum are viscous and relatively difficult to measure accurately. Third, blood, serum, plasma, and urine tend to foam while mixing, which can make volume adjustment difficult. Fourth, stored plasma may contain precipitated proteins, while stored urine may contain precipitated salts; both of these may block pipets.

Automatic pipets used in dilutions should be maintained and calibrated according to the laboratory SOP. Usually delivering of 98-102% of the nominal volume and 2% RSD are acceptable for automated pipetting devices; calibration should be performed every 3-6 months or more frequently when required (e.g. after maintenance, repair, etc).

There are two schools of thought for preparation of calibration standards. The first requires spiking of small volumes of standards on each working day using freshly prepared or diluted spiking solutions; such standards are not stored and are used up on the day of preparation, this approach seems to be favored by the Washington Conference Report [11]. The second approach permits spiking of standards in bulk, separating these into aliquots and storing under the same conditions as the QCs and study samples. In both situations the QCs are also spiked in bulk, separated into aliquots, and stored along with study samples. The argument in favor of the first approach is that by always using new standards the stability of samples is monitored, and that calibration standards are distinct from QCs. The argument for the second approach is that this procedure is acceptable if positive stability data are available, it is easier and more productive, and it avoids additional potential for bias by spiking once only. Additionally, the difference between calibration standards and QCs is that a calibration curve is forced through the standards, but not the QCs. This author prefers the second approach.

It is recommended to have between five and eight non-zero calibration standards, with single or replicate samples, in each analytical batch. More standards are needed for non-linear calibration curves. One calibration standard should be at the lower limit of quantitation (LLOQ); it is good practice to have the second standard at 2xLLOQ to define well the lower end of the curve, with all the other standards spread over the remaining range of concentrations (see more on LLOQ and ULOQ in sec. 10.3.2). It is recommended to have 3 standards per decade of concentration. Thus, a calibration curve over 1 - 200 ng/mL would have 8 non-zero standards, but a calibration curve over the range 0.1 - 100 ng/mL may require 10 standards. Drug-free matrix (blank) and drug-free matrix with the internal standard added (standard zero) should be a part of every analytical batch.

One of the frequently discussed issues is whether a single or duplicate calibration curve should be analyzed. Only minor gains in precision are achieved by increasing the number of measurements from one to two (see Section 2.2.2.1). Hence, a balance of costs and benefits is necessary. A typical batch size for LC and GC analysis is approximately 100 samples. A typical batch size is greater in immunoassays or analysis by LC/MS. Thus, standards, QCs, blanks and reference solutions analyzed in duplicate may constitute up to 30% of the entire run. This percentage decreases to approximately 18 % by the use of a single calibration curve. The price to pay for the single-curve approach is a greater probability for a run to be rejected due to loss of an LLOQ, ULOQ, or having a contaminated and unacceptable blank. That risk could be minimized by a skillful placement of calibration standards or, as some laboratories do, by measuring the LLOQ and ULOQ in duplicate or triplicate. While this approach virtually assures success, it is open to criticism as not all standards are treated the same way and may invite a creative interpretation of acceptance criteria. All things considered, it is best in the opinion of this author to measure calibration standards in duplicate.

Quality control samples should be prepared at 3 concentration levels and measured in duplicate with each analytical batch. One set of QCs should be 2-3 times the concentration of the LOQ. The second set should be at approximately 40-60% of the ULOQ, and the third at 70-90% of the ULOQ. QCs should be spread evenly throughout the analytical batch.

Assay accuracy is expressed as a percentage of the true value (μ), which is calculated according to the formula:

$$\text{Accuracy} = \left(\frac{\bar{x}}{\mu}\right) \cdot 100\% \tag{10.2}$$

where $\bar{x}$ is the mean of the observed value. The term "recovery" is sometimes used to describe accuracy; this usage should be discouraged because it is incorrect. The true value is assumed to be the nominal value at which a sample has been spiked; the accuracy can be expressed as percent of nominal. Some bioanalysts prefer to use observed value for the QCs instead of the nominal values [14]. The way to do it is to assay duplicate sets of QCs on 3 or 4 occasions, and calculate the interpolated mean concentration using n of 6 or 8. So established, the QCs' concentrations should remain in force for the duration of the project or until depletion.

The use of samples from dosed subjects was suggested for the assessment of precision [11]. The rationale for this approach is that despite the best efforts of the analyst it is virtually impossible to mimic a clinical sample by simple addition of standards to an appropriate medium. For example, a clinical sample may contain drug metabolites, concomitant medication and its metabolites as well as endogenous substances which may be impacted upon by the drug administration. Hence, pooled clinical samples could be used as an extra QC sample. Such a solution is often impractical for the reason of availability, yet may be very useful in some situations. For example, drug conjugates (glucuronides and sulfates) are notoriously difficult to obtain. A solution to this is to chemically or enzymatically hydrolyze these conjugates and measure the concentration of the free drug. However, a hydrolysis control is needed to make sure that such a process remains reproducible; a pooled subject sample could play this role [15].

Precision is a measure of the repeatability of a method and can be expressed by the relative standard deviation (RSD) as described in equation 2.6; this value is also commonly known as the coefficient of variation (CV). In today's bioanalytical assays an RSD of 5% or less characterizes a very precise method. An RSD of 5-10% is probably more common and represents an accepted industrial standard in terms of precision, while an RSD value of 15-20% suggests either a method of extreme difficulty and unusually low LLOQ or some analytical problems; such an RSD may be acceptable only around the LLOQ.

The ultimate goal of any method is to measure accurately and precisely the concentration of the analyte(s) in clinical samples; QCs are the best representation of clinical samples. Hence, accuracy and precision of a method should be estimated using the % nominal and RSDs calculated for QCs [8], and not back-calculated (interpolated) from the of calibration standards. A calibration curve is forced through the calibration points, and accuracy and precision based on standards always is better than those based on the QCs. On the other hand, back-calculated standards are useful and necessary tools in evaluation and adherence of the system to the selected mathematical model. Within-run precision and accuracy are evaluated during the validation experiments with a minimum of five replicate samples independent of standards at concentrations representative for the assay; a separate set of QCs could be used for that purpose.

### 10.3.2 Limit of quantitation

Limit of quantitation (LOQ) is frequently confused with sensitivity, it is particularly disappointing that this confusion was perpetuated in the Conference Report. Various ways of determining of the LOQ are presented in Chapter 2. The Conference Report recommends a very pragmatical approach to the LOQ. It is simply such a concentration which provides an RSD $\leq 20\%$ , and accuracy between 80 and 120%. The way to establish this LLOQ experimentally is to prepare at least five samples independent of standards at the concentration of the projected LLOQ, another set of five at concentration 2xLLOQ, one more at 4xLLOQ etc. These samples should be analyzed with a calibration curve. The concentration which fits into specification should be considered LLOQ, and the lowest calibration standard should be set at this value. The conference also endorsed other approaches to LLOQ, and alternative models of limit of detection are presented in Chapter 2 and references [16-19].

Although no mention of signal to noise ratio is made in the Conference Report [11], it is quite impossible to obtain acceptable precision and accuracy if the signal to noise ratio is smaller than 4:1. Introduction of new techniques and instrumentation improves the situation only temporarily. Most of the scientists working in the area of biological trace analysis are under a constant pressure to improve sensitivity and lower the LOQs; the question "Can you get lower than this ?" is proverbial. A chemical or instrumental breakthrough answers today's questions, and more insight into the nature of things invites more questions.

It should be noted that there is not only a lower limit of quantitation (LLOQ), but also an upper limit of quantitation - ULOQ. There are several reasons for existence of ULOQ: above a certain concentration a calibration curve may no longer be described by the chosen mathematical model, large chromatographic peaks may be truncated if a detector cannot handle an excess of signal, chromatographic peaks can be deformed by over-loading of the system, or a method simply has not been validated above a certain concentration. How to handle results that are above ULOQ (AQL) is discussed in section 10.4.4.

### 10.3.3 Specificity

There are two components of specificity. First, a bioanalyst must prove that other elements of the matrix do not generate (or significantly contribute to) the measured signal, and that the signal (chromatographic peak) is indeed generated by the analyte of interest. In chromatographic methods with detectors other than MS, an analyte is identified solely on the basis of its retention time. Characteristic UV spectra, excitation and emission fluorescence spectra or characteristic potentials in electrochemical detectors do not provide unequivocal proof of structural identity. Only MS/MS, and to some lesser degree MS, provide virtual certainty, that the signal observed was generated by the analyte of interest. One may hope that proliferation of MS instrumentation and smaller costs of these will render this paragraph obsolete in the near future.

Six samples of drug-free matrix obtained from six individuals should be used to prove lack of significant interference with the analyte of interest. The same biological matrix should be used for validation as that of the clinical/animal samples. In the cases of blood, plasma, serum or urine from humans or large animals the availability of drug-free biofluids does not present a problem. On the other hand matrices, such as cerebro-spinal fluid, bone marrow, sputum, bile, or fluids from small animals, may not be available in sufficient volume. In this case a surrogate matrix can be used instead for calibration standards and QCs preparation.

A couple of issues require elaboration. There is no such thing as "no peak". If one amplifies electronically the baseline in the area of interest then oscillations of the baseline and minor spikes will become visible and there is always some signal that could be integrated. The issue is how significant the contribution of the interference is allowed to be. At most the interference should be smaller than 33.3 % of the peak corresponding to LLOQ. This requirement has to be specified in the SOP. The interesting question is what to do if a calibration curve is assayed in duplicate and contains 4 drug-free samples ( 2 blanks and 2 standards zero), and one of them shows an interference at 50% of LLOQ, but not the other three. Should the batch be rejected on the strength of this one contaminated sample, or accepted on the strength of the mean value, which in this case would be 12.5% ? Situations like this should be anticipated and policies developed. (In this author's opinion the use of the mean value is acceptable).

On the one hand some projects involve dosing healthy volunteers whose diets are controlled and whose samples are relatively free of interference. On the other hand, in phase II of drug development a drug is administered to patients who routinely take concomitant medications and whose general health may be poor. Thus biological samples from patients are usually more difficult to analyze by chromatography, because drug-free patient samples tend to have more interferences. It is prudent and, in fact, necessary to test whether or not metabolites of the parent drug, as well as caffeine, common pain killers, and concomitant medication could interfere with the assay. This task may be difficult, as metabolites are not always available or even well characterized.

In order to prove that the substance being quantitated is the analyte of interest in assays other than MS, one has to inject a reference solution containing only the intended analytes (drug, and/or metabolite, internal standard) in solutions at the beginning and the end of the batch, and compare the retention times with those in the biological samples. The best solvent for these experiments in LC is usually the mobile phase. Too strong a solvent in liquid chromatography (for example, methanolic solutions injected at volumes greater than 10-20 μL into a typical reverse phase system) will provide a much shorter retention time and a distorted peak. In gas chromatography, injection of simple reference solutions may not provide peaks at all if there are active sites in the system which adsorb analytes. The way to avoid this is to mix a blank extract with a reference solution or by including a "carrier" substance in large excess in the reference solutions, which would not interfere with the assay, but saturate the active sites. If the sample preparation involves a back-extraction from ethyl ether or ethyl acetate or other relatively water-soluble organic liquids, the solvent in the reference solution should also be saturated with this reagent; otherwise the retention times will differ.

### 10.3.4 Recovery

Recovery understood as extraction efficiency is not mentioned in the Conference Report [11], but only in the DIA Workshop document [12]. Only very uncomplicated samples containing readily detectable analytes in a protein-free matrix (such as cerebrospinal fluid, urine, and saliva) can be injected directly onto a column. Most biological samples have to be processed on- or off-line in some fashion before entering a chromatographic column. There are no formal requirements regarding how high the recovery should be. The bioanalyst always tries to develop a method with recovery as close to 100% as possible. However, a recovery of 50% or less is generally considered acceptable, if it provides precise and accurate results and it is the best that could be achieved under the circumstances. Lower recoveries are frequently associated with poorer reproducibility, and warn an analyst or a reviewer to watch for unexpected problems or outliers. As bioanalysts are more often than not forced to work at the limits of sensitivity of the system, they can hardly afford a poor extraction efficiency, which in turn could decrease the LLOQ of the assay.

The absolute recovery can be calculated by comparing extracted QCs at three concentration levels, each prepared in replicates of five or six, with an unextracted calibration curve which has been prepared by mixing pure stock solutions at concentrations representing a 100% recovery and incorporating all the material losses due to the volume transfer. During the recovery study all the volume transfer should be done quantitatively unless an internal standard is used.

An issue affecting recovery is the so called "matrix effect". Sometimes the same concentration of analyte in different batches of biological matrix can yield different responses. A matrix effect is not always easy to discern since calibration standards and QCs are spiked into the same pool of matrix, and the concentration of the analyte in clinical samples is unknown. A matrix effect can be suspected, if a) a new batch of QCs respiked into a new pool of matrix does not fit into the calibration curve as defined by old calibration standards, b) study samples show very different concentrations than expected, or c) absolute peak heights (areas) of an internal standard are different in the calibrators, QCs, and study samples. Poorly validated methods are prone to this effect. A method can be tested for matrix effects by spiking a constant concentration of analyte into samples obtained from 5-6 different individuals, and assaying these samples in replicate along with the regular calibration curve.

The following are examples of matrix effects experienced in the laboratories of this author.

**a. Lipids in plasma samples** A solid phase extraction method [20] performed well until post-prandial clinical samples containing more lipids than usual were analyzed. The suspicion was confirmed by mixing pre-prandial samples with vegetable oil before extraction and observing a decrease in recovery. A weaker rinsing solution (less organic solvent) helped eliminate the problem.

**b. Sample pH** A liquid extraction was performed on plasma samples without buffering them as the physiological pH of plasma is 7.4, which was the appropriate pH for that extraction. However, stored plasma releases carbon dioxide and changes pH; thus a freeze-thawed plasma may reach pH of 8.5. Blood also may be affected by the same process but to a lesser extent.

**c. Ionic strength effects on ion exchange chromatography** Urine samples were injected directly into a column switching system containing an ion-exchange column [21]. Some samples provided suspicious results. It was discovered that these were very concentrated urine samples with much greater salt concentration. The volume of injection was reduced by factor of a 10 and the matrix effect disappeared.

**d. Protein content** Recovery following protein precipitation is frequently incomplete; solubility of the drug and its protein binding play a role. Total concentration of protein in normal human serum varies between 58 and 77 mg/mL depending on age and gender [22], and may be less in undernourished and sick individuals. In one experiment, recovery of triamterene from serum was measured at 50 ng/mL following precipitation with 10% perchloric acid. The serum was diluted with 0.9% sodium chloride solution in the following ratios (v/v): undiluted, 2:1, 1:1, and 1:2. The recoveries were 64, 75, 80, and 88%, respectively. An ideal internal standard could compensate for the recovery problems, but the errors might go undetected without it.

Hemolysis has been reported to influence the recovery and precision of data [23]. Wide variation in endogenous components from one species to another may also cause a matrix effect and necessitating re-development and re-validation of a method [10].

### 10.3.5 Response function

The theoretical basis for the establishment of the appropriate response function between the measured signal and the analyte concentration is described in Chapter 2. The Consensus Reports specify that a calibration curve should contain five to eight calibration standards (or more, depending on the nature of the function) with single or duplicate standards, should cover the whole intended range of concentrations, and be continuous and reproducible. Additionally, the simplest response function should be selected, the fit should be tested statistically, and an appropriate algorithm or graph presented.

What this means in practical terms is that during validation and/or the actual study the response function selected should remain constant, and not be changed from run to run. The common practice of splitting a calibration curve into two ranges for high and low concentrations should also be abandoned. The consensus reports lean very heavily towards the use of the simplest response function, i.e. linear calibration curve. However, this emphasis on linearity may cause problems. A subjective judgment whether or not a set of points represents a linear model may be at variance with statistical tests [24]. Thus a linear calibration curve may be forced on data that are slightly, but nevertheless clearly non-linear.

There could be several causes of non-linearity in chromatographic assays; receptor binding assays are non-linear by nature (see Chapter 2). Certain kinds of detectors provide non-linear responses, like the electron capture detector in gas chromatography, some older types of fluorescence detectors, or in fact any fluorescence or electrochemical detector if the calibration curve range covers concentrations of several orders of magnitude. To check the detector linearity one needs to inject increasing amounts of an unextracted analyte solutions and record responses. The analytical process may be also responsible for non-linearity, due to variable extraction recovery (see 10.3.6) or adsorption. To detect and document non-linearity one may use a number of techniques [25-27]:

- Visual assessment - subjective and requires an expertise in analytical methodology,

- Conventional analysis stemming from least squares regression - several approaches can be used like components of variance, lack-of-fit testing, quadratic regression, etc.

- Analysis of consecutive differences - simulates the visual assessment of linearity.

- Comparison of observed values against expected results (residuals, see 2.2.2.1)

Another simple test for linearity based on residuals is called "sign test" [24]. The signs of residuals should be distributed at random between plus and minus, if no systematic error is involved. In a sequence of signs (e.g.---++++++---), a curvature of the regression line could be suspected and a lack of linear fit.

This author finds particularly useful as a diagnostic tool the sensitivity plot [26], or rather a variation of it. Peak height (area) ratio or absolute peak height (area) divided by the nominal concentration gives a value which is called a "response factor" or "unit ratio". Assuming a zero intercept this value represents the slope of calibration curve at this point, and should be constant and equal to the overall slope of the linear calibration curve. If a decreasing/increasing trend in the value is observed, the response function is not linear. (Additionally, if response factors are constant over the whole calibration curve with the exception of the lowest standards, an interference hidden under of the analyte should be suspected).

Consider the authentic data presented in Tables 10.1 -10.3. Briefly, drug GG211 was extracted by protein precipitation from whole blood, the supernatant injected directly onto a reverse phase chromatographic column, and the drug detected by fluorescence detector after passing through a photochemical reactor, where its fluorescence was enhanced by UV irradiation. Absolute peak height (no internal standard) was used for quantitation.

Visual inspection of the graph in Figure 10.1 did not reveal curvature. The problem with this data set is that the calibration curve covers a wide range of concentrations and it is difficult to see the points with values which are close to the origin.

**Table 10.1**
Summary of representative experimental data

| Standard concentration (ng/mL) | Peak height (mV) | Response factor |
|---|---|---|
| 0.15 | 154 | 1027 |
| 0.30 | 289 | 963 |
| 1.00 | 1076 | 1076 |
| 6.00 | 7609 | 1268 |
| 30.00 | 41116 | 1371 |
| 60.00 | 83055 | 1384 |
| 100.00 | 138144 | 1381 |
| 100.00 | 141888 | 1419 |
| 60.00 | 86257 | 1438 |
| 30.00 | 41160 | 1372 |
| 6.00 | 7739 | 1290 |
| 1.00 | 1126 | 1126 |
| 0.30 | 345 | 1150 |
| 0.15 | 157 | 1047 |

**Table 10.2**
Various regression models as applied to data from Table 10.1

| Standard (ng/mL) | Residuals (% Deviation from nominal) | | | | | | | | | | | |
|---|---|---|---|---|---|---|---|---|---|---|---|---|
| | Linear (a) | | | Quadratic (b) | | | Ln transformed quadratic (c) | | | Power (d) | | |
| | e | f | g | e | f | g | e | f | g | e | f | g |
| 0.15 | 139.3 | 21.2 | 4.5 | 139.1 | 17.5 | 2.5 | 0.9 | -0.8 | -0.9 | -21.8 | -10.6 | -0.1 |
| 0.30 | 51.7 | -7.2 | -13.7 | 51.6 | -8.2 | -12.2 | -8.7 | -8.5 | -8.2 | -27.1 | -17.7 | -9.1 |
| 1.00 | 1.5 | -15.8 | -14.6 | 1.5 | -14.6 | -9.8 | -5.2 | -2.6 | -2.3 | -19.6 | -11.7 | -4.8 |
| 6.00 | -5.6 | -8.0 | -3.4 | -5.6 | -5.9 | 2.4 | 1.2 | 3.4 | 1.8 | -7.0 | -1.9 | 1.9 |
| 30.00 | -1.7 | -1.6 | 3.8 | -1.7 | -0.1 | 5.5 | 1.1 | -1.0 | -5.4 | -1.2 | 0.7 | 1.4 |
| 60.00 | -1.1 | -0.8 | 4.8 | -1.1 | -0.3 | 1.4 | -0.9 | -5.7 | -11.4 | -0.9 | -0.4 | -1.0 |
| 100.00 | -1.5 | -1.0 | 4.6 | -1.5 | -1.8 | -4.3 | -3.2 | -10.0 | -16.5 | -1.6 | -2.2 | -3.6 |

| | | | | | | | | | | | | |
|---|---|---|---|---|---|---|---|---|---|---|---|---|
| 100.00 | 1.2 | 1.7 | 7.4 | 1.2 | 0.7 | -2.0 | -0.6 | -7.8 | -14.5 | 1.1 | 0.4 | -1.2 |
| 60.00 | 2.7 | 3.1 | 8.8 | 2.7 | 3.5 | 4.9 | 2.8 | -2.4 | -8.3 | 2.9 | 3.3 | 2.6 |
| 30.00 | -1.6 | -1.5 | 3.9 | -1.5 | 0.0 | 5.6 | 1.2 | -0.9 | -5.4 | -1.1 | 0.8 | 1.5 |
| 6.00 | -4.1 | -6.5 | -1.7 | -4.1 | -4.3 | 4.1 | 2.8 | 5.0 | 3.4 | -5.4 | -0.3 | 3.6 |
| 1.00 | 5.1 | -12.2 | -10.8 | 5.0 | -10.9 | -5.8 | -1.0 | 1.8 | 2.0 | -15.9 | -7.7 | -0.6 |
| 0.30 | 65.0 | 6.2 | 0.4 | 64.9 | 5.5 | 3.0 | 7.9 | 8.7 | 9.0 | -13.1 | -2.3 | 7.5 |
| 0.15 | 140.8 | 22.6 | 6.0 | 140.5 | 18.9 | 4.2 | 2.7 | 1.1 | 1.0 | -20.3 | -8.9 | 1.7 |
| Sum of residuals | 422.9 | 109.4 | 88.4 | 422.0 | 92.2 | 67.7 | 40.2 | 59.7 | 90.1 | 139.0 | 68.9 | 40.6 |

y = absolute peak height, x = nominal concentration

a - linear regression $y = a + bx$; b - quadratic regression $y = a + bx + cx^2$

c - ln transformed quadratic $\ln(y) = a + b\ln x + c\ln x^2$ d - power $y = bx^a$

weighting factors: e - none; f - $1/x$; g - $1/x^2$

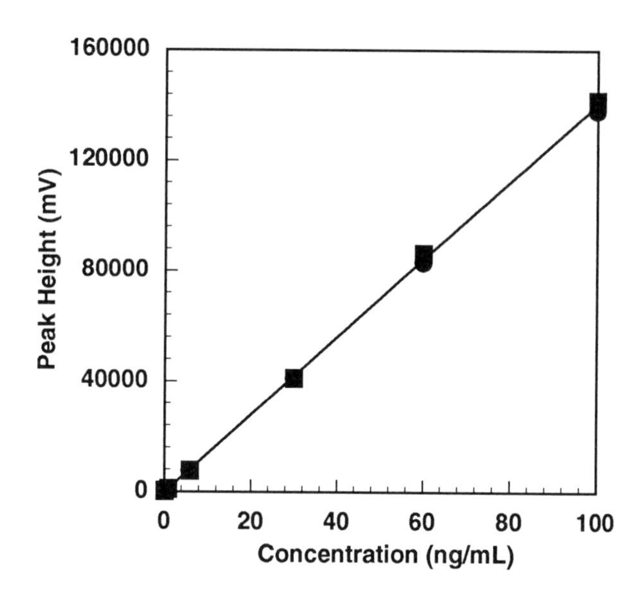


**Figure 10.1**
Linear calibration curve - graphical representation of the data from Table 10.1

The F-test for lack of fit requires repeat observations, which is another advantage of assaying the calibration curve in duplicate. The error sum of squares (SSE) is decomposed into two components called the pure error and lack of fit component. For details the interested reader is referred to the handbook by Neter et al [29]. The lack of fit test did not detect non-linearity; in this data set the test was not sufficiently sensitive to detect a deviation at low concentrations if the high concentrations follow the linear model.

However, the response factors calculated and presented in Table 10.1 show that the response factor at lower concentrations is approximately 1000 and it increases with concentration until it reaches 1400 at the highest concentration. Such a calibration curve cannot be linear.

**Table 10.3**
Calibration curve coefficients and correlation coefficients in various calibration algorithms as applied to data in Table 10.1

| Response function | Weighting factor | Correlation coefficient | a* | b* | c* |
|---|---|---|---|---|---|
| Linear | none | 0.99979 | -350.6 | 1405.5 | |
| | 1/x | 0.99970 | -99.80 | 1396.6 | |
| | $1/x^2$ | 0.99587 | -53.17 | 1321.8 | |
| Quadratic | none | 0.99979 | -350.0 | 1405.4 | 0.0010 |
| | 1/x | 0.99975 | -85.70 | 1360.2 | 0.4870 |
| | $1/x^2$ | 0.99779 | -35.29 | 1230.3 | 2.228 |
| Ln trans-formed quadratic | none | 0.99987 | 7.037 | 1.056 | 0.0015 |
| | 1/x | 0.99819 | 7.008 | 1.046 | 0.0060 |
| | $1/x^2$ | 0.99440 | 7.006 | 1.050 | 0.0093 |
| Power | none | 0.99987 | 1.010 | 1340.7 | |
| | 1/x | 0.99981 | 1.031 | 1223.2 | |
| | $1/x^2$ | 0.99931 | 1.051 | 1133.6 | |

*linear regression $y = a + bx$; quadratic regression $y = a + bx + cx^2$
ln transformed quadratic $\ln(y) = a + b\ln x + c\ln x^2$; power $y = bx^a$

The data from Table 10.1 were processed using several regressions, and sums of residuals were calculated for each equation. The results are presented in Table 10.2. The best models were the ln transformed quadratic equation without weighting factor, and power curve with weighting factor $1/x^2$. Several other functions provided acceptable results, including the linear regression with $1/x^2$ as the weighting factor. However, the sum of residuals in this model was approximately twice as large, indicating a poorer fit of the data.

Table 10.3 proves the fact well known to experienced bioanalysts that the correlation coefficient is a rather poor predictor of the goodness of fit. All 12 correlation coefficients are quite high and greater than 0.99, which is frequently used as an acceptance criterion. The unweighted linear equation provides a correlation coefficient of 0.99979; yet the results are totally unacceptable! There is some practical use for correlation coefficient in the sense that a high correlation coefficient does not ensure a good calibration curve, but a low one , e.q. <0.99, indicates that the calibration curve is biased with serious errors and is probably unacceptable.

Another issue related to the response function is the question whether to use peak height or peak area. From a theoretical stand point only the peak area is proportional to the mass of the analyte, and the peak height is related to the mass only as the height of a triangle, which approximates an ideal peak. Peak heights are much easier to measure manually, which is not so important now as it used to be, and also are easier to integrate in a crowded area, with partially overlapping peaks. Peak areas on the other hand are safer in that they are much less sensitive to changes in retention due to temperature fluctuation or decreasing resolving power of a column, as well as due to overloading of the system. The alternative use of peak height or area is worth trying at the validation stage in order to find an optimal response function and range of calibration curve.

### 10.3.6 Ruggedness (robustness) of a method

The Consensus Reports require a method to be reproducible and repeatable. Briefly, repeatability is precision achieved in the same laboratory by the same operator using the same equipment, while reproducibility is precision in different laboratories by different operators [30]. What the Consensus Reports do not mention is ruggedness or robustness[1] of a method, which is an important parameter. Validation should be performed using the same number of samples per batch as in the study, which is close to one hundred in most cases. The reason is to ensure that an appropriate precision and accuracy is obtained by an operator (human or robot) when challenged with a large number of samples, as well as to see if the system (chromatographic, robotic, etc) performs reliably over the long period of time (15-30 hours) that may be needed to complete an analytical run. The validation exercise should be limited to a certain number of runs. For example, if four acceptable runs are needed to complete the validation, no more than six attempts should be allowed. Should the six attempts fail to provide four acceptable results, the method should be sent back for re-development.

The bioanalyst always needs to do a balance of costs and benefits depending on physico-chemical properties of the molecule, concentrations required, and time considerations. The simplest solutions are quick, but not necessarily robust. For example, the order of extraction techniques from the biological matrix according to increasing difficulty and time consumption may be as follows: direct injection, protein precipitation, simple solid phase extraction, single liquid extraction, extraction followed by back-extraction; but the order in terms of the robustness of the chromatographic system may be reversed.

### 10.3.7 Stability

The most common reasons for instability of drugs in biological matrices are chemical, enzymatic, and photochemical processes [31]. The chemical reasons include hydrolysis of esters (e.q. diltiazem, aspirin), opening of the lactam ring (β-lactam antibiotics), opening of the lactone ring (campthotecin analogues), oxidation (phenols and naphthols), oxidation, dimerization and side-reactions (captopril). The enzymatic reasons for instability include hydrolysis of esters like procaine, esmolol and remifentanil by esterases. Finally, the light sensitivity affects for example nitrofurantoin, clomiphene, retinoids and fluoroquinolones.

---

[1] Editors note: in bioanalysis robustness and ruggedness are used interchangeably, which is in contrast to methods for the analysis of drug substances and dosage forms.

Stability has to be tested in the relevant matrix, under conditions encountered during execution of a bioanalytical study and includes bench stability, freeze-thaw stability, and long term storage stability. Samples from HIV positive patients are deactivated by heating at 56° C for 3-5 hours, hence this kind of stability testing should also be included in the validation exercise, if applicable.

On the bench stability testing simulates situations encountered during the sample collection and analytical work up, where samples typically remain at room temperature for 3-6 hours. At the end of that period stability samples should be compared with freshly thawed (or freshly prepared) control samples.

The freeze-thaw test mimics the situation where samples undergo freezing and thawing either during sample transportation, inventory, aliquoting or repeat analysis. The Conference Report recommends passing samples through two freeze-thaw cycles. These cycles should be at least 12 hours apart, if they are to simulate a real life situation. After the third thawing, the samples should be compared with a control set which has been thawed once. The freeze-thaw stability testing is of special importance in immunoassays, as the process of freezing and thawing causes precipitation of particulate matter, which may adsorb the analyte.

Long term storage stability testing should be performed over a period of time that equals or exceeds the time likely to expire in a typical project between the date of first sample collection and the date of the last analysis. Further evaluations may be made after 6 and 12 months of storage at the same temperature and in the same containers (geometry, caps) as the study samples. The last stage of this test should be a comparison of stability samples with control samples, which are presumed to be unchanged.

The Conference Report advises the use of two concentration levels (presumably high and low) in duplicate to evaluate stability, and does not offer any guidance regarding statistical calculations and acceptance criteria. The usual test for statistical significance between the stability and control samples is not useful, as the difference may be statistically significant, but irrelevant from a practical stand point. What is relevant or not is totally arbitrary; it is accepted that losses due to instability greater than 10% are relevant. Tim et al. [32] proposed a confidence limit based approach, which became a starting point for similar approaches. In their concept a compound is considered stable only if a relevant (>10%) degradation can be excluded with a probability of 95%. The imprecision of the measurements is taken into account and a distinction is made between significant and relevant stability. The authors propose to analyze 5 stability and 5 control samples at two concentration levels. This approach works well, if the RSD of the method is kept below 10%, which is achieved by analyzing concentrations higher than 5xLLOQ.

To overcome this shortcoming, Hooper et al [33] developed a similar method combining evaluation of ratios of means with a confidence limits approach. The confidence limits approach takes into account the variability of the analytical method. Approximately 10 replicates are used for both stability and control samples, at each concentration (i.e. 40 samples total). The differences in responses (not concentrations) between the replicates are assessed for the low and high concentrations combined, as well as separately to detect any relevant concentration-related effect. Another novelty of this approach was the use of stored control samples frozen in liquid nitrogen at -160° C. The frequent problem in stability testing is that control samples have to be prepared freshly; if there is any error in their preparation, it will be seen as instability or lack of it, if the bias compensates instability. To eliminate this uncertainty Hooper proposes to prepare both stability and control samples at the same time, store stability samples at the recommended temperature of -70° C or -30° C, while the control samples remain at -160° C.

Even if stability of a drug in un-adulterated matrix is insufficient, a bioanalyst can take preventive action to ensure sample integrity. An antioxidant like ascorbic acid or bisulfite can be added to avoid oxidation. Sample pH can be lowered by addition of citric acid to avoid hydrolysis. Esterases can be inactivated by addition of fluorides, EDTA, or dichlorvos. Addition of a chemical derivatizating reagent can also yield a stable entity. Lowering of temperatures is a good general way to slow down degradation. Samples can be either flash-frozen immediately after collection or kept in icy water. During processing samples may be kept at 0-4° C and processed quickly. Refrigerated autosamplers are readily available to ensure stability of extracts. If samples cannot be stored for any period of time, they may have to be analyzed immediately at the collection site [34]. If freeze-thaw is the problem, the samples can be divided into a number of tubes, and any repeats done using only new tubes.

One unusual consequence of instability is when a metabolite or degradant produces the parent drug by undergoing a chemical reaction during the analytical process [35,36]. In this particular case the instability will result in an apparent increase in the parent-drug concentration.

## 10.4 Application of a validated method to routine drug analysis

### 10.4.1 Organization of analytical batch

In most studies, all clinical samples from one subject should be analyzed in the same run to avoid between-batch variability, which tends to be greater than within-batch variability (see 2.2.1). This is of particular importance in bioequivalence studies.

The analytical batch should be started by injection of a reference solution followed by crucial samples (LLOQ, blank, ULOQ), to have an early indication of whether or not the run is under control.

There are no general rules for the placement of calibration standards in the batch. Some analysts prefer to disperse them evenly throughout the whole batch while others start a batch with the first calibration curve and end a batch with a second. QCs should be spread evenly throughout the analytical batch.

### 10.4.2 Acceptance Criteria

The same criteria for acceptance of an analytical batch should be used for the method validation and execution of a routine project (within-study validation). According to the Conference Report, agreement of back-calculated QC concentrations with nominal concentrations is a sole acceptance criterion, which is set at +/- 20%. This accuracy should translate into precision and RSD no greater than 15%, and 20% around the LLOQ. Four out of six QCs must fulfill this condition, at least one at each concentration level.

In industrial practice somewhat stricter criteria are generally used. Very common are acceptance criteria set at 10,15, and 20% (from highest QC concentration to the lowest), or 10,10, and 15%, or any similar combination. Additionally, a set of secondary acceptance rules are used. These deal with:

- deviations of standards from the nominal concentration; no greater than 10,15, or 20%.

- limiting the number of calibration standards that can be rejected from a regression (too great a number of rejected standards suggests lack of robustness or other analytical problems)

- required correlation coefficient of at least 0.99

- lack of interferences in drug-free samples (blank, standard zero)

Less common are some other rules, like even distribution of positive and negative bias, rejecting values between unacceptable QCs [37], and inclusion of assay-specific secondary quality controls.

In the opinion of this author and others, this element of the consensus reports invites criticism and needs improvement. Hartmann et al [13] calculated that in order to obtain mean values within the limits of +/- 15% with probability of 95% the bias and RSD should be 8% with n=5. The fixed range approach accepted by the Conference is totally pragmatic, and not based on statistics. It describes quite well the current state of bioanalysis that the authors of the report and the industry at large were comfortable with. The fixed range criteria confuses precision and accuracy and at the same time creates irritating contradictions summarized in Table 10.4.

If the acceptance criteria are set at 10, 15, and 20%, then batch A will be accepted and batch B rejected although batch B seems to be much better than A, with total sum of deviation at 46.6% as compared to 171.9%.

**Table 10.4**
Quality control data according to the fixed range acceptance criteria

| QCs Conc. (ng/mL) | Batch A | | | Batch B | | |
|---|---|---|---|---|---|---|
| | Back calculated | Deviation (%) | Decision | Back calculated | Deviation (%) | Decision |
| 3.00 | 2.41 | -19.3 | accepted | 2.39 | -20.3 | rejected |
| 3.00 | 5.12 | 70.7 | rejected | 3.01 | 0.3 | accepted |
| 30.00 | 34.45 | 14.8 | accepted | 34.56 | 15.2 | rejected |
| 30.00 | 25.55 | -14.8 | accepted | 30.11 | 0.4 | accepted |
| 75.00 | 82.33 | 9.8 | accepted | 74.77 | -0.3 | accepted |
| 75.00 | 43.11 | -42.5 | rejected | 67.42 | -10.1 | rejected |

A good set of acceptance criteria should be scientifically valid, able to detect errors and false alarms, easy to use and provide immediate answers. The Conference Report recognized that a confidence-interval approach is an acceptable alternative for acceptance criteria.

#### 10.4.2.1 The 99% Confidence interval

The analyst observes measurements which are related to concentration, not the concentration itself. These measurements provide only a probability that the true concentration will be within a certain range [38]. If analytical errors are random they follow the Gaussian normal distribution of population. Hence, 68% of the results fall within one standard deviation of the mean, 95% within 1.96 (popular 2) of the mean, and 99.7% within 3.09 (popular 3) of the mean.

**Table 10.5**
Quality control data according to the 99% confidence interval acceptance criteria

| QCs Conc. (ng/mL) | Batch A | | | Batch B | | |
|---|---|---|---|---|---|---|
| | Back calculated | Deviation (%) | Decision | Back calculated | Deviation (%) | Decision |
| 3.00 | 2.41 | -19.3 | accepted | 2.39 | -20.3 | accepted |
| 3.00 | 5.12 | 70.7 | rejected | 3.01 | 0.3 | accepted |
| 30.00 | 34.45 | 14.8 | accepted | 34.56 | 15.2 | accepted |
| 30.00 | 25.55 | -14.8 | accepted | 30.11 | 0.4 | accepted |
| 75.00 | 82.33 | 9.8 | accepted | 74.77 | -0.3 | accepted |
| 75.00 | 43.11 | -42.5 | rejected | 67.42 | -10.1 | accepted |

The 99% confidence interval is equal to

$$99\%CI = \bar{x} \pm 2.58s \tag{10.3}$$

where s is the standard deviation. Gross errors (e.g. bad chromatography or sample processing, instrumental problems) should be eliminated from calculations. The acceptance criterion is simple: all QCs must fall within the confidence interval [37]. Let us see how the 99% confidence interval works in the case from Table 10.4. The standard deviations at 3.00, 30.00, and 75.00 are 0.309, 2.156, and 3.755, respectively. The 99% confidence intervals are respectively 2.20 - 3.80, 24.44 - 35.56, and 65.31 - 84.69 (Table 10.5).

Using this approach, batch A will be rejected, while batch B should be accepted. The confidence interval based acceptance criteria are easy to use and provide an immediate answer, though do not address accuracy. However, they may be even more liberal than the fixed range. At low concentrations the RSD is commonly at the range of 10 - 15%, and the acceptance criteria will be +/- 25.8 - 38.7%.

#### 10.4.2.2 Statistical process control

The ultimate objective of a process is to make products which conform to pre-determined specifications. A control chart is a graphic comparison of process performance data to computed "control limits" drawn as limit lines on a chart. This branch of statistics was started in 1931 by an American industrial scientist, Walter Shewhart. He introduced the idea that a statistical process control (SPC) chart could be used to detect, as early as possible, any deterioration in the product being manufactured. In other words, whether or not the process is in control. However, a state of statistical control merely means that only random causes are present and it does not necessarily mean that the result meets specifications.

There is a multitude of SPC charts: charts of ranges, averages, SD, moving ranges, cumulative sums and so on, as well as a wealth of literature describing them [e.g. 39,40]. For most control charts, the control limits are calculated on the basis of the average +/- 3 times the standard deviation. If n is less than 10, then the range is used instead of SD; as this is the case of bioanalysis.

A simple procedure for chart of averages and chart of ranges is presented in Table 10.6. (For simplicity, the actual charts are not shown). The mean of the averages is calculated, as well as the mean of the ranges. Factors A2, D3 and D4 are taken from Table 10.7 for n=2. For a batch to be accepted all points must be within the control limits. Additionally, a batch must be rejected if 2 out of 3 successive points are at 2 SD (if SD is used) or beyond, 4 out of 5 successive points at 1 SD or beyond, or 8 successive points on one side of the center line.

Again, the batch A should be rejected by the chart of ranges as the range for QC 75.00 is 39.22, which is greater than upper control limit of 39.08. The batch B should be accepted.

**Table 10.6**
Quality control data according to the average chart

| QCs Conc. (ng/mL) | Batch A | | | Batch B | | |
|---|---|---|---|---|---|---|
| | Back calcul-ated | Devi-ation (%) | Decision | Back calcul-ated | Devi-ation (%) | Decision |
| 3.00 | 2.41 | 3.77 | uncertain | 2.39 | 2.70 | accepted |
| 3.00 | 5.12 | | | 3.01 | | |
| 30.00 | 34.45 | 30.00 | accepted | 34.56 | 32.34 | accepted |
| 30.00 | 25.55 | | | 30.11 | | |
| 75.00 | 82.33 | 62.72 | rejected | 74.77 | 71.10 | accepted |
| 75.00 | 43.11 | | | 67.42 | | |

**Table 10.7**
Control limit factors (excerpted from ref [38])

| n | A2 | D3 | D4 |
|---|---|---|---|
| 2 | 1.880 | 0 | 3.268 |
| 3 | 1.023 | 0 | 2.574 |
| 4 | 0.729 | 0 | 2.282 |
| 5 | 0.577 | 0 | 2.114 |

These control charts take into account the precision and accuracy of an assay. However, if one uses additional or extended acceptance criteria, then no immediate answer is possible, as even results generated many batches later may influence acceptance or rejection of the first batch. Of course, these additional rules intend to detect trends in the process.

### 10.4.3 Dilutions

There are 3 reasons to dilute a sample in bioanalysis: to place sample concentrations above the quantitation limit within the calibration range (see sec. 10.4.4), to study parallelism in RIA, and to analyze samples with insufficient volume. In these cases the dilution should be done using the same matrix as the original sample, even if blank matrix from the same individual may not be available. One has to be careful with dilutions of samples of concentrations close to the LLOQ as the diluted samples may be classified as below the quantitation limit (BQL). For example, if LLOQ is 1 ng/mL, the dilution factor is 2, and the back-calculated concentration (no dilution factor included) is 0.77 ng/mL, the reported concentration should be BQL, and not 1.54 ng/mL.

### 10.4.4 Repeats

Every bioanalytical laboratory should develop an SOP for repeat assays. The policy defined in the SOP must be implemented before starting a study. Most examples requiring reassay result from some kind of analytical or technical difficulty:

a. **Bad chromatography** includes interfering peaks making the integration impossible, no peaks, radical differences in chromatographic pattern, chromatographic column failure etc.

b. **Lost sample** broken sample tubes, leaking pipet tips, leaking screw caps, etc.

c. **Bad processing** human or robotic errors like reagent or internal standard omission, adding excess of reagent or internal standard, etc.

These errors should result in a documented audit trail of deficient chromatograms, notes to the study file listing lost samples or errors in the processing, or computer print-outs in case of robotic systems. No numerical results are associated with failed analysis and samples should be repeated as single samples.

Occasionally, clinical samples exhibit concentrations above the validated range (AQL = above quantitation limit). Such samples should be diluted with the same matrix and repeated singly. If a concentration obtained is only slightly above the highest standard, the Conference Report recommends either redefining the calibration curve range by adding a higher standard or analyzing diluted sample. The DIA Consensus permits an extrapolation above the top of the calibration curve limited to one standard deviation or 15%.

Study samples sometimes provide results which formally and chromatographically look correct, yet seriously contradict previous results. The goal of a bioanalyst or pharmacokinetist is to provide results for which there is a scientific basis. At the same time, it is appropriate to challenge suspected results. One may suspect a pharmacokinetic outlier if a pre-dose sample from naive subjects contains a measurable drug concentration, if a profile exhibits a halved or doubled maximum contrary to known pharmacokinetics, or if concentrations are very different (500-1000%) than expected. Such samples, which could be called "suspected outliers (SO)" provide numerical values and repeats should be done in duplicate. A bioanalytical laboratory should also develop a decision tree dictating a verdict in every foreseeable case to eliminate arbitrary decisions. A very good decision tree has been developed by Lang and Bolton [3]. Briefly, a 15% agreement between data is considered a confirmation if the repeats are done in duplicate, or 30% if only one repeat was possible. If results are too far apart, no result is reported (NR).

Suspected pharmacokinetic outliers should be evaluated and excluded using a well known outlier test. In bioanalysis, the volume of the study samples are generally limited and samples can rarely be assayed with sufficient number of replicates for any meaningful outlier test to be applied. It should be also noted that occasional outliers do not influence the outcome of a study if correct numbers of subjects are selected to ensure appropriate statistical power .

Repeat assays are required as a matter of policy by Canadian Health Protection Branch, or by some clients in the contract industry to verify correctness of the data. These repeats should flagged as "confirmation points" or "client-requested. "

Finally, if samples are repeated in error, these results cannot be ignored and must be evaluated according to the appropriate decision tree.

### 10.4.5 Reporting of data

Both the initial and within-study validation should result in a report describing the procedure, its performance, and study results, where applicable. While the formats found throughout the industry differ in scope and detail, the report should contain certain essential elements. The data should be presented in tabulated form and include:

**a. Table of calibration curve parameters including slope, intercept, and correlation coefficient** All parameters should contain a sufficient number of significant figures to calculate the concentrations accurately. Slope of the calibration curve may change from day to day, yet it remains a valuable diagnostic tool. A consistent slope suggests a robust assay. Dramatic changes may suggest instability, method modification, errors, or detector maintenance (new lamp, polished electrodes etc). Intercept values consistently above zero may suggest an interference under the peak of interest.

**b. Table of back-calculated (interpolated) standards** The table should be complete, with no empty spaces. If a standard has been rejected, its value should be provided in parentheses or flagged and an explanation provided. Samples lost or disqualified for any reason should be flagged. Intra-day precision and accuracy should be calculated providing the mean, standard deviation, RSDs, % of nominal and number of observations. A sufficient number of significant figures should be provided for reviewer to verify calculations and arrive at the same conclusions.

**c. Table of back-calculated quality controls** All the rules specified above apply to this table as well. Additionally, all evaluable QCs have to be reported and included in the statistics, whether or not these QCs meet the acceptance criteria. One may apply an appropriate outlier test and exclude some aberrant QC values from statistical calculations; these values should be clearly marked and the outlier test either referred to or described. Precision and accuracy calculated on QC data is the precision and accuracy of the method.

Another point of concern is the way QCs are usually presented and statistical parameters calculated. As QCs are assayed in duplicate global statistics (all results included) are usually calculated at the end of a project. Such statistics are neither within-run nor between-run, although they are typically presented as the latter. The correct way of presenting the between-run precision and accuracy is to split the QC data into separate columns (first low QC, second low QC, first middle QC ... etc.) and calculate the mean, standard deviation, RSD and % nominal separately. One may then characterize a method by providing either a range of values for the between-run precision and accuracy, or the global statistics. Another approach to the same problem is to analyze the data by ANOVA, which permits the simultaneous calculation of the global, within-run and between run RSDs.

The bioanalytical and/or validation report should include an unequivocally identified procedure in the form of a validation report, SOP, or a method sheet, list of relevant SOPs, acceptance criteria, and analytical time lines. The report could include a list of deviations from the approved protocol, special cases, description of a typical analytical batch organization, and specimen handling.

## 10.5 Validation of Immunoassays

The Conference Report treats the microbiological methods and immunoassays equally. However microbiological methods are generally not recommended for bioavailability, bioequivalence and pharmacokinetic studies [41]. Principles of analytical validation for chemical methods apply also to immunoassays. There are method-specific issues which can be addressed separately.

### 10.5.1 Specificity

In contrast with chromatographic or spectroscopic methods immunoassays (IA) do not leave a comprehensive, auditable record which can be reviewed and critically evaluated. Hence, ensuring the specificity in IA is of critical importance. Specificity can be compromised by metabolites, concomitant medication and/or its metabolites, and endogenous matrix components. The specificity can be proven by any of the following techniques or combination:

**a. Cross-reactivity** Comparison of curve displacement can provide indication of cross-reactivity. The response in the presence and the absence of a metabolite or concomitant medication should be evaluated. Often the lack of metabolites can make this test impractical.

**b. Comparison of calibration curves prepared in buffer and matrix** If there is no difference between these two curves, the matrix does not contribute to the response and the endogenous compounds do not cross-react. A potential matrix effect should be further examined and excluded by investigating blanks and analyte spiked into matrices obtained from at least 6 individual and various donors, such as different gender and social groups.

**c. Parallelism** Three elements are evaluated in a parallelism study: the amount of cross-reactivity between the antibody and related compounds, binding affinity of the ligand in both standards and clinical samples, and potential matrix effects. Clinical samples and calibration standards analyzed undiluted and diluted with the same matrix provide results which linearized should yield two superimposable or approximate parallel lines.

**d. Separation** Specificity can be tested by comparing untreated samples with those that were extracted or separated by HPLC. If an antibody is not specific, samples should be pre-treated by extraction or chromatography before undergoing an immunological reaction.

### 10.5.2 Quantitation

Calibration curves for immunoassays should be prepared in the same matrix as the study samples. Samples and standards should be treated in an identical fashion. The acceptance criteria for precision and accuracy are the same as for chemical assays. If appropriate precision and accuracy is difficult to achieve, duplicate samples should be assayed, which is normal practice for immunoassays.

Immunoassay calibration curves are non-linear and require a greater number of calibration standards. Contrary to the chromatographic methods the best precision and accuracy is not achieved at high concentrations. In fact, high concentration samples may need to be diluted to ensure that the concentration falls within the calibrated range. The LLOQ and ULOQ is defined by assaying replicate samples which can yield appropriate precision and accuracy.

The shape of the response function in IA is frequently hyperbolic or sigmoidal (see Chapter 2). The most common curve-fitting models are [42,43]:

- point-to-point linear interpolation
- polynomial equation
- spline functions
- semilogarithmic plot
- reciprocal plot
- log-log plot
- hyperbolic function
- log-logit model
- four parameter logistic
- five parameter logistic

The Conference Report mentions two other issues related to immunoassays: use of commercial kits and measurement of metabolites. A more elaborate scheme to evaluate and validate RIA kits was developed by the World Health Organization Expert Committee on Biological Standardization [44], which states that the validation should include, where possible:

- a concentration-response curve representative of the assay

- statistical estimation of the mean, standard deviation, and range of percent bound/maximum binding (B/Bo), slope, and intercept for calibration curves from several assay runs
- an antibody dilution curve
- cross-reactivity data (relative $ED_{50}$s for closely related compounds likely to cross react)
- demonstration of parallelism over two or more orders of magnitude
- results of accuracy experiments showing the ability to obtain 100% accuracy over the working range of the assay
- Scatchard plots and equilibrium constants
- a measure of within- and between-assay variability at three concentration levels
- a time course of the RIA reaction for one or more temperatures
- comparison with other analytical methods
- detailed information on all parameters likely to affect assay performance (e.g., pH, ionic strength, additives, incubation time and temperature, separation technique)
- references to the scientific literature
- studies to substantiate the satisfactory performance of the assay
- range of normal and pathological values

This scheme is a good example of a thorough and detailed validation exercise of a receptor-binding assay either commercially available or developed in-house.

## 10.6 Validation, re-validation, and cross-validation

There are at least four types of validation experiments in bioanalysis:

**a. The original validation** (before-study, immediately after method development); described in sec. 10.3

**b. Within-study validation** described in sec. 10.4

**c. Cross-validation** which should be performed when a new individual is introduced to the method, the sample matrix is changed, the sample volume is changed with no relative change in assay LLOQ (for example, sample volume decreases by factor of 2 and LLOQ increases by 2), the calibration range is moderately extended, robotic system replaces manual processing, or when a significant period of time (3 months, for example) has elapsed from the last use of the method. In other words, cross-validation involves an unchanged method that will be used by different personnel under different circumstances and there is a reasonable assumption that the method would work as well as before. The process should involve fewer batches than full validation (two as compared to four) and may not require a written report at the end.

**d. Re-validation** should be performed when modifications introduced to the original method no longer can be recognized as minor, or a method is transferred to another facility which does not have previous experience with this particular method. Re-validation should be done in the same format as the original validation and should result in a validation report or an amendment to the original validation report.

The question of whether changes and modifications to an original method are major or minor is difficult to answer. Analytical methods may improve with use, new devices and techniques can replace old ones, and sometimes changes are forced by manufacturers of reagents and equipment, who may discontinue production, change product specification and process, or go out of business. Policies regarding changes in a validated method could vary from conservative, demanding re-validation after a single small change, to liberal, permitting use of the method as long as the main concept of the method is the same. One has also to keep in mind the costs of validation, which today for a simple method may amount to

$5,000 - 10,000, and reach $20,000 for a LC/MS or GC/MS method. Below are a couple of examples which in opinion of this author do not constitute a major change:

- change of chromatographic column (the same type, different manufacturers)
- modification of mobile phase by moderate increase/ decrease of organic content, addition of modifier, moderate change of pH
- changes in volumes of reagents
- changes in injection volume and flow rate
- minor corrections of wavelength (2-5 nm) or temperature

A single change does not require a validation, but the sum of all these may warrant it; the final decision should be left to the discretion of an analytical director. A conscientious analyst should test specificity (lack of interference in the blanks) after changing any part of a chromatographic system or extraction process.

Some examples of changes that in the opinion of this author require re-validation are as follows:

- change of detection system (e.g. from UV to fluorescence detector)
- replacement of the liquid phase extraction with the solid phase extraction or *vice versa*
- change, addition, or elimination of internal standard (historical data may exist and be usable)
- drastic changes in chromatography (i.e., from reversed phase to normal phase chromatography, gradient instead of isocratic)
- LLOQ is decreased
- range of calibration curve is drastically increased by one order of magnitude or more

- response function is changed (historical data may exist and be used

A validated method may be altered for a certain purpose or inadvertently. In any case, the change should be described in a note to the study file, and its potential impact evaluated. Modifications for this purpose should be authorized by the unit's analytical director and a rationale provided in writing before implementation.

## Acknowledgement

The author thanks Christine Grosse, Julie Tomlinson, Glenn Smith and Ian Davis for assistance in editing this chapter.

## References

1. R. J. N. Tanner, *Methodol. Surv. Biochem. Anal.*, **20**, 57-64 (1990)

2. D. Dadgar and M.R. Smyth, *Trends Anal. Chem.*, **5**, 115-117 (1986)

3. J. R. Lang and S. Bolton, *J. Pharm. Biomed. Anal.*, **9**, 357-361 (1991)

4. J. R. Lang and S. Bolton, *J. Pharm. Biomed. Anal.*, **9**, 435-442 (1991)

5. A. G. Causey, H. M. Hill and L. J. Phillips, *J. Pharm. Biomed. Anal.*, **8**, 625-628 (1990)

6. H. T. Karnes, G. Shiu and V. P. Shah, *Pharm. Res.*, **8**, 421-426 (1991)

7. L. J. Phillips, J. Alexander and H. M. Hill, *Methodol. Surv. Biochem. Anal.*, **20**, 23-36 (1990)

8. D. Dell, *Methodol. Surv. Biochem. Anal.*, **20**, 9-20 (1990)

9. J. A. F. de Silva, *Methodol. Surv. Biochem. Anal.*, **10**, 298-310 (1981)

10. G. S. Land and R. D. McDowall, *Methodol. Surv. Biochem. Anal.*, **20**, 49-56 (1990)

11. V. P. Shah, K. K. Midha, S. Dighe, I. J. McGilveray, J. P. Skelly, A. Yacobi, T. Layloff, C. T. Viswanathan, C. E. Cook, R.D. McDowall, K. A. Pittman and S. Spector, *J. Pharm. Sci.*, **81**, 309-312 (1992)

12. A.C. Cartwright, *Drug Inf. J.*, **25**, 471-482 (1991)

13. C. Hartmann, D. L. Massart and R.D. McDowall, *J. Pharm. Biomed. Anal.*, **12**, 1337-1343 (1994)

14. D. Dell, *Methodol. Surv. Biochem. Anal.*, **20**, 64 (1990)

15. K. Selinger, H. M. Hill, D. Matheou and L. Dehelean, *J. Chromatogr.*, 493, 230-238 (1989)

16. H. Keiser, *Two Papers on the Limit of Detection of a Complete Analytical Procedure*, Hafner, New York (1969)

17. M. Uhlein, *Chromatographia*, **12**, 408-411 (1979)

18. J. McAinsh, R. A. Ferguson and B. F. Holmes, *Methodol. Surv. Biochem. Anal.*, **10**, 311-319 (1981)

19. M. A. Curtis, Proceedings of AAPS Southeast Regional Meeting, Research Triangle Park, NC, p.10 (1993)

20. H. M. Hill, I. Smith, I. Fairbrother, K. Tennant and J. Robson, *Methodol. Surv. Biochem. Anal.*, **23**, 327-334 (1994)

21. D. M. Morris and K. Selinger, *J. Pharm. Biomed. Anal.*, **12**, 255-264 (1994)

22. C. Lentner ed., Geigy Scientific Tables, Vol.3, 8th Edition, Ciba-Geigy Ltd., Basle, Switzerland (1984)

23. P. Heizman and R. Gora, *Methodol. Surv. Biochem. Anal.*, **23**, 365-372 (1994)

24. M. Thompson, *Analyst*, **107**, 1169-1180 (1982)

25. D. W. Tholen, *Arch. Pathol. Lab. Med.*, **116**, 746-756 (1992)

26. H. T. Karnes and C. March, *J. Pharm. Biomed. Anal.*, **9**, 911-918 (1991)

27. J. S. Krouwer and B. Schlain, *Clin. Chem.*, **39**, 1689-1693 (1993)

28. National Committee for Clinical Laboratory Standards. NCCLS EP6-P Evaluation of the Linearity of Quantitative Analytical Methods. Philadelphia, PA, National Committee for Clinical Laboratory Standards, 1986

29. J. Neter, W. Wasserman and M. H. Kutner, *Applied Linear Statistical Models*, 2nd edition, p.123-130, Irwin, Homewood, Illinois 1985.

30. International Organization for Standardization, in Accuracy (Trueness and Precision) of Measurement Methods and Results, ISO/DIS 5725-1 and 5725-3, (Draft versions 1990/91)

31. P. Heizman, K. Zinapold and R. Geshke, *Methodol. Surv. Biochem. Anal.*, **23**, 351-357 (1994)

32. U. Timm, M. Wall and D. Dell, *J. Pharm. Sci.,* **74**, 972-977 (1985)

33. W. Hooper, D. Lessard, K. Selinger, R. Parkes and D. Dadgar, *Pharm. Res.*, **10**, S-46 (1993)

34. D. O. Scott, D. S. Bindra and V. J. Stella, *Pharm. Res.*, **10**, 1451-1457 (1993)

35. J. A. Jersey, S. A. Guyan and I. M. Davis, *Pharm. Res.*,**11**, S-58 (1994)

36. H. M. Hill, I. Smith, I. Fairbrother and K. Tennant, *Methodol. Surv. Biochem. Anal.*, **23**, 359-364 (1994)

37. H. T. Karnes and C. March, *Pharm. Res.*, **10**, 1420-1426 (1993)

38. M. Thompson and R. J. Howarth, *Analyst,* **105**, 1188-1195 (1980)

39. J. M. Juran ed., *Quality Control Handbook*, 3rd edition, McGraw-Hill (1979)

40. G. B. Wetherill and D. W. Brown, *Statistical Process Control*, Chapman and Hall, London, New York, Tokyo, Melbourne, Madras (1991)

41. T. M. Ludden, at Open Session on Validation and FDA Requirements for Analytical Work for Bioavailability and Bioequivalence Studies, AAPS Annual Meeting, San Diego, CA (1994)

42. J. R. George, D. F. Palmer, J. J. Cavallaro and W. M. Wagner, *Principles of Radioimmunoassay*, U.S. Department of Health and Human Services, Atlanta, GA (1984)

43. C. P. Price and D. J. Newman ed., *Principles and Practice of Immunoassay*, Stockton Press, NY (1991)

44. WHO Expert Committee on Biological Standardization, 26th Report, World Health Organization Technical Report Series No. 565, Geneva (1973)

# Chapter 11

# Analytical Methods for Cleaning Procedures

*Thomas M. Rossi and Ralph R. Ryall*

## 11.1 Introduction

A key step in the validation of a pharmaceutical manufacturing process is the validation of the equipment cleaning procedure. Ensuring the quality, purity and safety of a given drug product requires that the manufacturer demonstrates that the cleaning procedure used will consistently minimize the probability of product adulteration. Possible sources of adulteration include previously manufactured product, cleaning agents and solvents, and the water used during the equipment cleaning cycle. Associated with this is the potential for microbiological contamination that may result from an inadequate cleaning process. In general, pharmaceutical manufacturers demonstrate the adequacy of an equipment cleaning procedure by conducting a systematic validation process that includes analysis for possible contaminants. As defined by Harder [1], a validated cleaning process is "a procedure whose effectiveness has been proven by a documented program providing a high degree of assurance that a specific cleaning procedure, when performed appropriately, will consistently clean a particular piece of equipment to a predetermined level of cleanliness."

## 11.2 Cleaning Procedures

Pharmaceutical equipment cleaning procedures are detailed in written documents to ensure that the equipment is cleaned in a consistent manner, according to a validated process. The level of detail in these cleaning protocols will vary depending on the complexity of the equipment and the nature of the product residue being removed from the equipment. Normally, the protocol will specify equipment disassembly requirements, the type and concentration of any cleaning agent(s) used, the volume of cleaning solution, the types of rinse solvents used, the number of rinses, and procedures for proper handling and disposal of rinse solutions. In addition, documentation should include procedures for the sampling and testing that are conducted to evaluate the levels of residual contaminants.

## 11.3 Sampling Plan

The integrity and safety of a pharmaceutical product can be easily jeopardized by the presence of even trace level contamination. Used alone, visual inspection for cleanliness, as evidenced by the lack of visible residues, often will not provide the necessary assurance that the probability of cross-contamination has been minimized. This is especially critical with those drug residues that have high toxicity and/or therapeutic potency. Accordingly, the validation of a given cleaning procedure should normally include sampling and testing of the cleaned surfaces to verify the adequacy of the process. The most common sampling procedures include: (1) swabbing of the equipment surfaces, and/or (2) sampling of the final rinse fluid. A third technique that is sometimes discussed but is not recommended [2] as the preferred choice is manufacture of a placebo batch in the cleaned equipment, and then testing for contamination in representative samples from this batch.

The validity of the sampling strategy is a key element of performing a meaningful cleaning validation test. Irrespective of the sampling technique chosen, it is essential to verify that the contaminant is thoroughly extracted from this sampled surface. Furthermore, it is equally important to ensure that the contaminant is fully recovered from the sampling medium (swab, solution, or placebo) prior to the analysis.

Swabbing of equipment surfaces is usually performed with some appropriate swab material to wipe a defined area that has been exposed to the product. Test results for the swabbed area are then used to calculate the level of cleanliness for the total surface area of the equipment. Suitable swab materials include cotton, paper, or other cellulosic material, and synthetic materials such as polyester. The swabs are used dry, moistened with water, or saturated with an appropriate solvent. Among the advantages of using the swab technique is that one can be more assured of recovering any surface residue that could not be adequately washed and/or rinsed from the area that is being wiped. In addition, it affords the ability to sample difficult-to-clean areas. The downside is that extrapolation of the test results on small, poorly accessible areas may transpose into exaggerated levels of residue for the total equipment surface.

The sampling of rinse fluid is operationally an easier technique. However, a major disadvantage is that the residue(s) may be insufficiently soluble in the rinsing fluid, thus giving erroneously low results. It is also quite possible that the rinse fluid will not come into contact with all of the equipment surfaces that are contacted by product during a manufacturing process. It follows, therefore, that combined testing of both swab samples and rinse fluid samples will provide the highest level of assurance that an equipment module has been adequately cleaned. The decision tree reproduced in Fig. 1 provides for a systematic means of simplifying the sampling procedure [3]. Through this controlled evaluation of both swab-sampling and rinse-sampling, one can determine if the simpler rinse-sampling technique used alone will provide assurance of an acceptable cleaning process. However, in order to shorten development times for cleaning validation assays, it may be preferable to utilize swab sampling by default and not investigate other options unless the swab test cannot be validated.

A third technique for sampling equipment for the presence of contaminants is to process a placebo batch using the equipment of interest and then testing the batch for contaminant levels. This technique is operationally a costly one and, as indicated in the U.S. Food and Drug Administration (FDA) Inspection Guide [2], there is no assurance that any contaminants will be uniformly distributed throughout the placebo batch. This procedure is also more demanding from an analytical testing point of view, since any low level contaminant must be determined in the presence of a complex matrix. This may require a more elaborate sample preparation scheme and/or the use of a more selective and sensitive analytical technique.

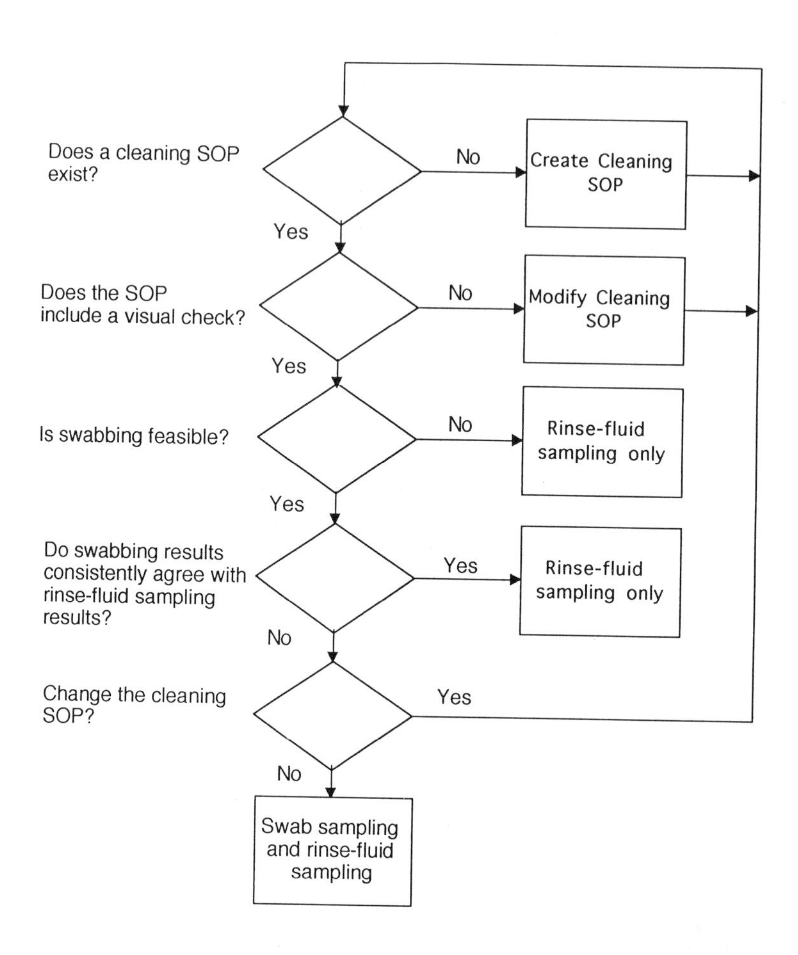


Figure 11.1. Decision tree for evaluating sampling methods (Adapted from Ref. [3])

## 11.4 Analytical Method Validation

Some essential considerations must be made in choosing a suitable analytical test method for testing completeness of cleaning. As a start, the quantitation limit of the test method must be lower than the predefined acceptance limit for residue levels. As will be discussed later in this chapter, the acceptance limit for a cleaning process is the maximum allowed level of contamination that has been judged not to jeopardize the safety, quality and integrity of a pharmaceutical product. Most state-of-the-art analytical techniques are capable of detecting and quantifying compounds at levels well below those which would be considered a safety concern. More often than not, a quantitation limit in the low parts-per-million range is adequate for the testing of rinse fluids and swab samples. With all of the advances in analytical technology, the task of detecting residual contaminants is often less formidable than the task of reproducibly cleaning the equipment to a defined tolerable limit of contamination.

Other considerations to make in choosing a suitable test method include the accuracy and precision achievable, as well as the specificity/selectivity that the method allows. The combined sampling and test method should be challenged for accuracy by testing the recovery of potential contaminants that have been added to the equipment surfaces or some small scale panel comprised of the same surface components. For example, a known amount of drug product or cleaning agent can be applied to a known surface area of equipment material and then rinsed or swabbed according to the sampling protocol. This should be done over a range of potential contamination levels, i.e., at the acceptance limit, as well as above and below the acceptance limit. The samples are then tested according to the proposed analysis method to ensure that the combined sampling and test method provides adequate recoveries. In the case of swab samples, it is usually necessary to test recovery of the potential contaminant from the swab by adding various amounts of contaminant to representative swabs and carrying them through the extraction procedure. These same experiments can be further refined to provide additional information regarding method precision, e.g., by having the experiments repeated on multiple days by more than one analyst.

### 11.4.1 Method Specificity/Selectivity

The issue regarding the desired level of test method specificity/selectivity is one that requires some case-by-case evaluation. In most instances, the key potential contaminant is considered to be the last active drug agent(s) that was in contact with the equipment surfaces. Typically, if it can be demonstrated that the active agent(s) has been removed to an acceptable residual level, and if the equipment is visibly cleaned, then the assumption is made that drug product excipients and other non-therapeutic agents have been adequately removed. For this reason, most test methods used to support cleaning validation are based on existing analytical assays for the active drug agent. Modifications are usually required to account for the lower levels of drug being assayed and for the different sample matrix involved.

It is important to consider that, in many cases, the test method does not have to be stability indicating. If the compound being assayed is stable in the sample matrix then it will not be necessary to demonstrate the ability to determine the compound in the presence of potential degradation products. Thus, it may be adequate to substitute a non-selective test, such as a UV absorption measurement, for a more time-consuming and resource-intensive method. Nevertheless, consideration should be given to other potential sources of interference. These may include other components of the product residue, i.e., product excipients, agents used in the cleaning process, and compounds that may be introduced by the sampling procedure or analytical sample workup. Interferences arising from the cleaning process include the components of any detergent as well as any acids, bases or organic solvents that may be used to clean the equipment. Likewise, one needs to consider potential interferences from the swabs that may be used in sampling and from any solvents used for extraction of the analyte.

The FDA considers the removal of detergent residues an important aspect of the overall cleaning process [2]. If detergents are used to clean the manufacturing equipment, it is expected that they will be completely removed by rinsing, and that the removal be validated by testing for absence of residues. Unlike the situation with product residues it is not always as straightforward in selecting which component of the detergent to target for residue analysis. The pharmaceutical manufacturer may not have access to the specific composition of the detergent, although suppliers are usually willing to provide information on the chemical class of active agent(s). Moreover, there is commonly more than one component that is responsible for the cleaning action of the detergent product. It then becomes a question of whether to develop a test for a specific target component or for the whole detergent product. If choosing a single component, one should focus on testing for the particular component that is the most difficult to rinse from the equipment surface. A systematic procedure for identifying the selective adsorber in formulated detergents was described recently by Smith [4]. He used FTIR spectroscopy to identify the component of a formulated cleaning agent that was the last to rinse from stainless steel and borosilicate glass surfaces.

One analytical technique that provides an excellent means of conducting whole detergent product analysis is that referred to as Total Organic Carbon (TOC) analysis. Using this technique, organic compounds are oxidized to $CO_2$ which is then quantitated, typically by non-dispersive infrared absorption or by conductivity. TOC analysis is non-specific and offers low detection limits, potentially down to low parts-per-billion levels. Furthermore, TOC analysis is theoretically capable of quantitating any carbon-containing compound. These attributes make TOC ideal for measuring low levels of residual detergent in water rinses and on swabs with low carbon background, thus precluding the need to investigate the complexity of the detergent formulation components. Because of its broad applicability to all organic carbon-containing compounds TOC analysis has been proposed as the method of choice in validating cleaning procedures for biopharmaceutical products produced by recombinant DNA technology [5]. Cleaning validation for this class of pharmaceutical products presents a unique challenge because of the wide range of potential contaminants, including proteins, carbohydrates, nucleic acids, as well as the cleaning detergents. However, FDA site inspectors generally give strong preference to product specific cleaning validation assays. This is particularly true in multiple product facilities, which are becoming more common for proteins as more recombinant products are gaining market approvals.

It is not always clear when it would be appropriate to use a non-specific analytical technique, such as TOC analysis, and when a specific assay for residual drug product contaminant should be used. FDA guidelines strongly imply that a direct measurement of the residue or contaminant should be made [2]. However, there needs to be further discussion between industry and the regulatory agencies to define how specific this measurement must be in order to ensure an acceptable level of cleaning. The efficiency advantages of TOC analysis, combined with the logical justification of using TOC results as a worst case assessment of residual levels, should make this technique more widely accepted from both industrial and regulatory perspectives than it currently is.

This discussion has focused on analysis for potential chemical residues. However, the potential for microbial contamination should also be considered in establishing a cleaning process for manufacturing equipment. Generally, pharmaceutical firms have established microbial standards for both sterile and non-sterile facilities, and these are routinely monitored to ensure conformance.

## 11.5 Acceptance Criteria

In establishing the suitability of an equipment cleaning procedure it is essential to define an acceptable maximum level of residue below which the equipment can be considered adequately cleaned. Today's analytical technology allows the detection and quantitation of chemical components at extremely low levels; determinations at the part-per-million and even the part-per-billion levels are quite common. Such low-level residues may often be below the cleaning capability of even the most thorough procedures. Therefore, establishing an acceptance criteria based solely on the detection limit of the analytical assay may be too stringent from a practical point of view, and it may be far more demanding than necessary to ensure the safety and integrity of any succeeding product manufactured in the equipment. For these reasons pharmaceutical manufacturers generally establish acceptance criteria based on the following considerations:

- the upper limit of allowable residues must be practical and reproducibly achievable by a reasonably designed cleaning procedure;
- the upper limit of allowable residues must be verifiable by analytical testing; and
- the upper limit must be judged safe and acceptable based on toxicological and pharmacological factors.

The latter consideration will involve input from toxicologists and pharmacologists on the lowest (threshold) dose of residue at which toxicity and/or therapeutic effects can be expected. Establishing this level in an R&D setting can be problematic because it is routinely necessary to manufacture clinical products before the toxicological and pharmacological activities of the investigational compound have been fully characterized. In such cases, it is sometimes useful to assume that the lowest planned clinical dose represents a no effect, or threshold, dose. It is common practice to attach a safety factor of at least 10x to this dose level, resulting in an upper limit that is some fraction, e.g. 0.1, of the threshold dose. When working with research materials during development, a higher safety factor is recommended. Traditionally, the manufacturer calculates for each piece of equipment the maximum tolerable residue level, assuming that the total residue can be carried over into the smallest batch size/maximum dose combination of the next product to be processed. This approach must be used with caution since several pieces of equipment in the manufacturing train, each with its own calculated threshold, could result in an intolerable accumulation of residue in the next product. Alternatively, it may be useful to establish the limit based on cleaning "efficiency". This approach requires that the residue levels are defined based on 99.X% removal of the active ingredient from the equipment. This essentially ensures that the next batch would contain no more than 1/99.X carryover from the most recent batch. A number of publications discuss in detail the different methods for calculating acceptance limits for contaminating residues [3, 6-7]. The general theme behind all of these methods is that the acceptance limits be based on sound scientific criteria.

Any procedure used for establishing acceptance criteria for cleaning validation should take into account the visual cleanliness of the equipment. The cleaned equipment must be visibly free of all residues regardless of the established safety threshold. Limits calculated on the basis of toxicity/pharmacological potency could easily exceed the levels which would give a visually detectable residue. This is particularly true for those drug products that have relatively safe toxico-pharmacological profiles. In those cases, common sense and good science would indicate that the equipment should be cleaned until no residues are visible.

## 11.6 Conclusions

Good analytical science is a vital factor in the establishment of a validated cleaning procedure for pharmaceutical manufacturing equipment. This is reflected both in the need to determine proper sampling techniques for trace levels of potential contaminants and in the choice of analytical test method that will provide adequate sensitivity and specificity. Proper choices on the part of the analytical chemist will help to ensure the timely throughput of samples while maintaining the safety and integrity of drug products that will be manufactured using the equipment under study.

Presently, all cleaning validation assays are performed in the QC laboratory by working on samples collected from the "cleaned" equipment. The development of in situ assay methods would reduce the complexity of validation of these methods by eliminating several mass transfer steps. For example, given adequate sensitivity and portability, techniques such as FT-IR reflection absorption spectrometry [4] could provide advantages for rapid analysis and high levels of product selectivity.

## References

1. S. W. Harder, *Pharm. Tech.*, **8(5)**, 29-34 (1984)

2. FDA Guide to Inspections of Validation of Cleaning Processes, July 1993

3. A. O. Zeller, *Pharm. Tech.*, **17(10)**, 70-80 (1993)

4. J. M. Smith, *Pharm. Tech.*, **17(6)**, 88-98 (1993)

5. R. Baffi, G. Dolch, R. Garnick, Y. F. Huang, B. Mar, D. Matsuhiro, B. Nepelt, C. Parra, and M. Stephan, *J. Parent. Sci. Tech.*, **45(1)**, 13-19, 1991

6. D. W. Mendenhall, *Drug Dev. Ind. Pharm.*, **15(13)**, 2105-2114 (1989)

7. G. L. Fourman and M. V. Mullen, *Pharm. Tech.*, **17(4)**, 54-60 (1993)

## Chapter 12

# Computer Systems and Computer-aided Validation

*Joseph G. Liscouski*

## 12.1 Introduction

Computer hardware and software have become fundamental tools in the conduct of laboratory work. Since the late 1970s, when the initial FDA regulatory requirements were published, the entire structure of the computer industry has changed. The tool of the specialist, highly trained computer expert, is now the commodity item; one can purchase a complete system with a laser printer from a wide choice of vendors in the same store you buy eggs and automobile tires. The nature of the typical end-user has changed from the technical applications specialist (predominantly laboratory and engineering) to anyone.

This change in emphasis has had its effect on software in particular. Operating systems (OS), once strongly designed for real-time work and revised in an evolutionary, continual improvement manner, are now geared toward the commercial user. Changes (upgrades[1], new utilities, etc.) are made in response to competitive

[1]upgrades were once a mechanism for fixing problems with a system or a controlled means of adding new features. Historically changing a number to the left of the decimal point in version numbers (going from version 1.0 to version 2.0) indicated a significant change while changes to the left of the decimal point indicated a minor change. That is no longer the case. The jump in Microsoft Windows from version 3.0 to 3.1 for example was a major change resulting from significant alterations in the OS and causing many commercial software packages to be modified to work properly. The next change - to version 4.0 - promises to be radical, eliminating the need for the underlying DOS software and developing incompatibilities with existing systems. This would not be as significant a problem as it is except for the fact that software vendors will drop support for earlier versions of an OS shortly after a new one is introduced.

pressures to gain more market share often without regard to the impact changes may have on existing users. In addition, the current software architectures are structured to permit new functions to be added by other software vendors through the use of plug-in modules. These modules may cause conflicts to occurs with other plug-ins or with other software packages.

Similar changes exist in computer hardware. There are a number of vendors of computer systems all claiming compatibility with Microsoft's Windows and DOS, Apple's Macintosh OS[2] and various versions of UNIX. Added to this mix is the relatively recent arrival of emulator boards and software that allow software designed for one operating system to run under another.

The benefits of these changes are clear: lower cost systems, a mind-boggling range of software offerings, and increasing power to do things easily. It is easy to take pieces of computer systems hardware and software and assemble them into a "system". Developing confidence that the assembly will do the things you want it to without unpleasant side-effects and interactions is not so easy, particularly in laboratory applications. That is why the validation process is critical to the success of modern computerized laboratory work.

Validation is a matter of establishing documented evidence, which provides a high degree of assurance that a specific process will consistently produce a product meeting its predetermined specifications and quality attributes [1, 2]. It is a means of building confidence into a system or process or an assurance that it does function in accordance with needs and specifications set out for it. The purpose of the material in this chapter is to provide a guide to the literature and practices of laboratory computer systems validation.

---

[2]Apple is in the process of licensing its System 7 to other hardware manufacturers on the PowerPC

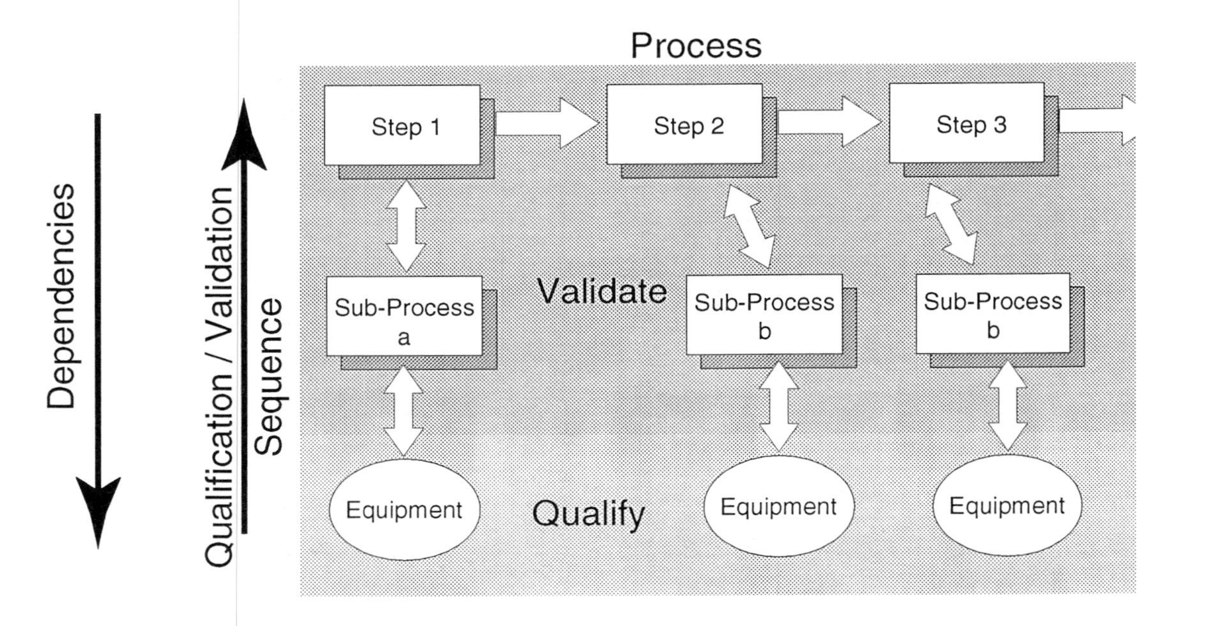


**Figure 12.1**
The relationship between processes, sub-processes, and equipment used in a process, showing the parts that require qualification and validation

We will also look at the direction that laboratory computing technology is taking, its impact on systems design and validation, and the possible problems that can occur as newer operating systems, tools and non-traditional programming environments develop. In doing this work we will make one departure from current practices. Laboratory computer/information systems (applications - both commercial and custom developed, OS, and hardware) are usually treated as a collection of independent entities. One may validate a Laboratory Information Management System (LIMS), or a data acquisition/analysis system, or a robotics system, each as an independent entity. The view of this author is that these should be treated as part of a laboratory-wide system and designed as such [3]. The intent of this chapter is to provide an overview of the considerations that should be taken into account in the development of validation programs for laboratory computer systems, networks, and automated equipment. Since each laboratory is responsible for designing its own program - work that cannot be delegated - we are not going to attempt to create a "10 steps to better validation" process.

## 12.2 Validation and Qualification, what do they mean when applied to computer systems

According to the FDA a process can be validated, and equipment used in that process should be qualified for use (Fig. 12.1) [1, 2, 4]. A process is a sequence of steps, that when followed yields an expected result; it may include sub-processes, some of which may be used in more than one step (common sub-processes only need to be validated once, but they have to shown as appropriate for each instance that they are used). Equipment includes physical things, tools, and components that are used to conduct that process. This may seem obvious, but when one looks at a computer systems (hardware and software) how does one apply the terms? A standard hand-held fixed-function calculator would have to be qualified for use in a process. A fixed-function non-programmable (no loadable software, just ROM code) data capture device may be qualified. The identical functionality in either example, implemented in a software application, layered on an operating system and hardware may require validation.

Automated systems and computers come in a variety of shapes

and forms, some based on conventional desktop packaging, others in vendor designed packages which may be an integral part of an instrument. Some guidelines are needed for what has to be qualified as equipment and what has to be validated. For the sake of this discussion, something will be considered as "equipment", requiring qualification, if its behavior is fixed regardless of its environment[3] (assuming it is operating within the specifications for use defined by the vendor). Anything else should be validated. This is not a trivial point, since it will help define the work that needs to be done in developing and executing the validation plan.

Consider the calculator noted two paragraphs above. It does a fixed number of tasks and always does them the same way regardless of where or how it is being used; as a piece of equipment, it would be qualified for use. The identical functionality, rendered in software running on a PC would have to be validated. The difference is due to the fact that the environment may vary depending upon the type of computer being used, modifications that may be made to an underlying mathematical package (this is becoming more of an issue as operating systems move to object-oriented implementations), the availability of hardware-assisted mathematics, etc. The vendor of the calculator program may require certain minimum configuration requirements, but provide some latitude in the actual operating environment (for example, specifying that at least an Intel 80386 be used, placing only a lower bound but not anticipating the future processors or emulators).

The distinction between what is and is not equipment, what requires validation or qualification, needs careful consideration as computer technology changes. A few years ago it was a simple matter: if you could touch it, it was hardware and thus "equipment". Software was viewed as intangible, as the set of instructions that made things actually function (disk, printers, graphics display, etc.). Computer systems had one CPU, the applications software and

[3]"Environment" should be taken to include the physical setting (temperature, atmosphere, vibration, humidity, etc.) and also anything that can effect its behavior. For computer systems this would include versions of the operating system, other applications that may be running at the same time a particular piece of software is being used, the hardware configuration (disk type, math libraries, presence of a floating-point math co-processor, etc.

operating system, and the peripheral hardware accessed and completely controlled by "drivers" or "device handlers" that were part of the OS. The situation today is not as simplistic. The typical desktop system is really a network of microprocessors, each with its own programming (contained in ROMs and firmware) and command structures that allow the components to communicate and work. As networked systems and client/server applications become developed, devices on the network that may have once been viewed a separate computer applications systems requiring validation (CPU[s], loadable software, etc.) will become networked boxes - equipment - that carryout a particular function. A file server for example, that contains its own CPU(s), ROM instructions, loadable (and upgradable) software and disk array, all acting as though they were one very large storage unit[4], acting as equipment requiring qualification. Today's Postscript laser printers are another example. They are attached and used as printing *equipment*, yet they contain CPUs, programming, disks [optional on some, internal on others] and communications protocols and connections, and may contain bugs requiring correction.

Equipment can be shown to be usable in a process if it can be qualified. That qualification requires a clear set of specifications for what it is supposed to do (based on the process that it is supporting), a means of determining if it meets those specifications, and the ability to maintain it. Demonstrating that it functions properly can be a matter of both user defined testing and a successful history of use in similar applications. Custom designed computer equipment, or other laboratory automation equipment, should follow a qualification process similar to that for custom software applications: functional and design specifications.

### 12.2.1 The Hierarchy of Software and Hardware Systems

There are three general categories of software all of which reside on the computer hardware: CPU(s), disks, and memory:

- **Application Specific Software**[5]: A program adapted or

[4] RAID storage, Redundant Array of Inexpensive Disks, is a good example of this type of system.

[5] Application Software: A program adapted or tailored to the specific user

tailored to the specific user requirements for the purpose of data collection, data manipulation, data archiving or process control.

- **General Purpose Software** - commercial software packages that are used as provided to do a specific function. These packages have wide distribution (tens of thousands to millions of copies) and include word processors, compilers, spreadsheets, etc. Note: Ref. 2 does not include this category, however the developments over the last seven years require it. During the 2nd Annual Conference on the Validation of Laboratory Systems [5] Ken Chapman described this type of software as COTS (Commercial Off The Shelf software). In the laboratory this would include *commercial* LIMS, data acquisition packages and others - in-house developed software would be treated as applications specific. Modifications to COTS software can put it into the Applications Specific catagory[6].

- **Operating System Software** - the underlying control software that manages the hardware, provides basic commands for using a computer (copying files, initiating program start-up, file maintenance, etc.) and defines the storage structure for the computer systems. Examples: DOS, Microsoft Windows, the Macintosh O/S, UNIX (and its variants), VMS.

According to the current FDA guidelines, only the applications

---

requirements for the purpose of data collection, data manipulation, data archiving or process control. - PMA Definitions from the Pharmaceutical Technology Conference 1991 [11]

[6] This concept is new to laboratory applications and needs to be considered carefully in validation programs. It is best, at least in the short term, to take a conservative approach and validate the software. Since the conference was only recently held, regulatory inspectors may not be aware of this idea, and may hold differing views. It needs to be tested in practice, before it is treated as an accurate reflection of the agreed upon state-of-the-art.

specific software needs to be validated. General purpose software, the O/S, and the hardware have to be qualified for use. While all of the software is key to the successful operation of a system in the lab, the second and third items on the list above, are exempt from validation due to wide-spread successful use. Note again, that care should be taken in qualification of those software items.

From mid-1994 through the first quarter of 1995, a number of reports have complicated the COTS concept. The most widely publicized was the Pentium FPU math errors, important not only because an error occurred, but because the vendor, Intel, chose to ignore the problem until faced with a severe public reaction. In addition, several income tax preparation packages were shipped with several flaws, and reports [6] indicate that the beta-test process used by vendors to find bugs in software may itself be flawed. The COTS concept is based on the assumptions that the production software is developed in an orderly process, well tested before release, and that volume production and wide-spread use lead to progressively better packages. That notion may have been true ten years ago, when software developed in an evolutionary process. In the mid-1990s, the evolutionary process has been greatly modified with wholesale technology shifts that are masked by a slow progression in version numbers. These problems are not limited to applications software but extend to the operating systems themselves. The final Windows-95 beta test software distribution reportedly[7] contains serious design flaws that may extend to the basic architecture of the operating system.

General purpose software is designed with certain assumptions (operating systems used, the version of that O/S, hardware requirements, support for networking, compatibility with other software, and, known interferences). The requirements at one level become part of the qualification/product selection criteria for the next

---

[7] "Win95 beta lays eggs" [7] - Note: At the time this material was prepared there was considerable discussion about the extent of the problem, the availability of software fixes to the problem, and the impact on the final product. Software development issues are not limited to Microsoft, a number of other vendors including IBM's OS/2 have their own issues to contend with. Microsoft's problems receive more attenetion and publicity than the others due to the company's current position in the computing marketplace.

lower level. One needs to pay attention to the literature, particularly the computer trade press where significant bugs, interferences, incompatibilities, and other short-comings will be published. In software it is better to assume that something is suspect unless it has been proven to work (including combinations of software packages, even if they will not be executing at the same time).

While this is true in all O/S systems, it is particularly of concern in the Microsoft Windows environment. There has been considerable concern in the press (*InfoWorld* and *PC Week*) about software installation procedures which modify files without notifying the user or providing installation logs. Removing applications is a particularly interesting problem since one is never quite sure that everything that been changed has been put back to its original condition. There are a growing number of products designed to solve this issue, but they are yet to be proven.

User programming built upon commercial packages (macros, the construction of forms) should be validated as applications software. That validation should be done in the same environment as would be used in production work. A change in the environment may be sufficient cause for re-validation. That covers most of the bases except for two: limited distribution commercial software and commercial software that is customized. Both these situations require validation. Low volume commercial packages may not have built up a sufficient history of use to demonstrate their reliability. Customized packages, either through custom programming or the inclusion/exclusion of options, effectively reduce the user community for a product to very few users, maybe just one. It should also be pointed out that the FDA's primary concern is with applications packages, particularly those that collect, analyze, report or manage laboratory data[8] . An internal FDA task force has been organized to address the issues of general purpose software.

This is far from the final word on the topic because it is a dynamic situation. The rate of development of operating systems, the modification of operating systems through the use of "plug-in" modules, improperly tested systems (vendors using the consumer as an extended beta test site), the layering of products, and the development of links between different applications and commercial

---

[8]Private communications from Paul Motise, FDA, May 25th, 1994

packages, can cause a re-evaluation of any of these points. If there is uncertainty, the conservative course of action is to validate rather than just qualify. The closer something is to the acquisition, analysis, and, management of laboratory data, the more likely a validation protocol is required.

## 12.3 Who has to Validate?

Validation is a necessary part of the design and implementation of automated systems and processes. The development and installation of automated equipment are engineering activities and the 'customer' should require a validated/qualified system as part of the 'customer acceptance criteria'. Everyone should validate/qualify each process, system and piece of equipment that is used in the laboratory, regardless of whether the industry is pharmaceutical, environmental, biotechnology, bulk and specialty chemical, clinical, or petrochemical. The validation process is enforced in regulated industries (FDA's Good Laboratory Practices [GLP], Current Good Manufacturing Practices [cGMP], the EPA's GLP and the proposed Good Automated Laboratory Practices [GALP] - currently treated as recommendations and still in draft form) and those interested in European markets (ISO 9000).

Validation is a process of developing proof that processes and the systems supporting them work. Manufacturing companies may see it as an expense, but in reality it is a matter of product improvement. In laboratory work the products are data, information, and knowledge. A validation program builds confidence that those products are high quality and that they can be taken as a sound basis for decision making. The effort also shows up as a reduction in expenses, reducing or eliminating the need to retest materials should questions about products occur. Repeated tests duplicate expenses, and delay current work; if problems are encountered overall confidence in data is reduced and larger programs of retesting may be required to either uncover problems or show that you are dealing with an isolated incident.

Tetzlaff - then at the FDA - in his May 1992 *Pharmaceutical Technology* article [8] states "There should be no question in anyone's mind that FDA expects computer systems to be validated. It also should be obvious that the cGMP sections covering laboratories (Parts 211.160 and 211.165) apply to automated systems." The GMPs in Parts 211.160 and 211.165 cover the key points. The question to be addressed next is how to go about it.

## 12.4 Overview of the Validation Process

Validation is a process that may be either prospective or retrospective[9] . Our primary concern is with the former. Planning for validation should be an intimate part of the design of any process. That effort involves clear definitions of what is to be accomplished, what is required to implement the process, the development of specifications for systems and the criteria for choosing one out of several possibilities and the means of showing that the choice does meet the criteria. Part of that work, when applied to computer systems, is the definition of the hardware and software needed.

Retrospective validation is a questionable activity when applied to computer systems, though in some cases it may be unavoidable. Those cases include the operating system software and commercial software with wide use and application (word processors, etc.). The questions arise when applied to laboratory applications software. Even though a software system may have been in use for a long period of time, there is always the possibility of encountering unexecuted code simply because the conditions necessary to execute it have not occurred (combinations of events, out-of-the-ordinary situations).

There is a fundamental distinction between validation and testing: the validation of a process is done within a context, testing is context free. In order for something to be validated, the first thing that has to be defined is its purpose. From that point, everything else follows including (but not limited to):

---

[9] Retrospective Validation (PMA CSVC): Establishing documented evidence that a system does what it purports to do based on an analysis of historical information. Authors note: retrospective validation is not an acceptable substitute for formal prospective validation programs. It should only be considered for remedial action in combination with prospective validation for a non-compliant establishment.

- definition of need
- product requirements/specifications
- product selection criteria (given a set of requirements, how do you choose between competing vendors, what are the mandatory requirements, which are optional)
- determining how well a product fits your needs, and what modifications may be needed to improve the fit
- determining the equipment needed to use a software product (type of computer, special hardware requirements, and so on)
- demonstrating that the system works (testing) and meets the original criteria
- providing documentation (Standard Operating Procedures, user manuals, etc.) and training people to use it
- providing ongoing maintenance (including periodic testing, upgrades, updating documentation etc.), ensuring that the system remains in a validated state

The validation process begins as soon as a need is determined - the documentation of that need is the first part of the process. Once begun, validation is a non-stop process passing through several phases until settling into a maintenance mode. There is an initial phase when a product is first selected and put into production. Then there are follow-up phases as modifications are made or adaptations to changes in process. Each phase should be fully documented (what was done, why, when, and by whom) including any re-validation depending upon the nature of the work.

### 12.4.1 The Process of Laboratory Work - Defining the Need

Figures 12.2 and 12.3 show the flow of work in a typical laboratory. The existence of a model such as this is a key element in the development of validation plans since it provides an overall

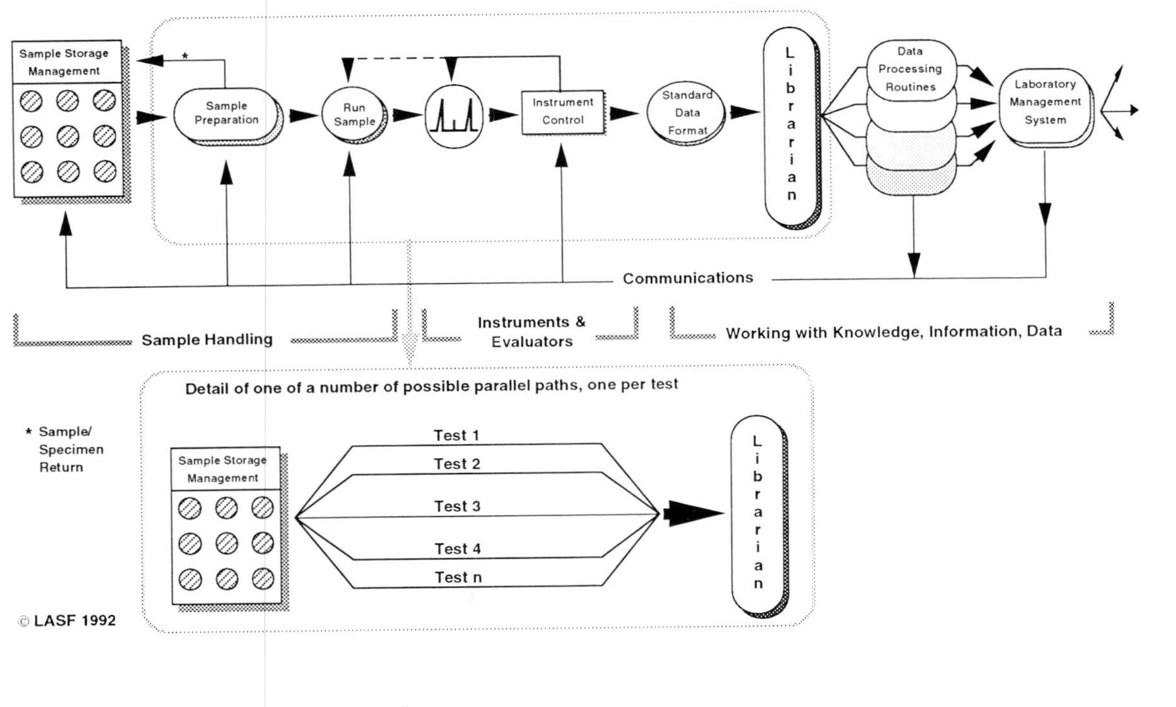


**Figure 12.2**
This figure illustrates the process nature of laboratory work and the relationships that must be included as part of a validation program. This figure is used with permission of the LASF (a non-profit foundation).

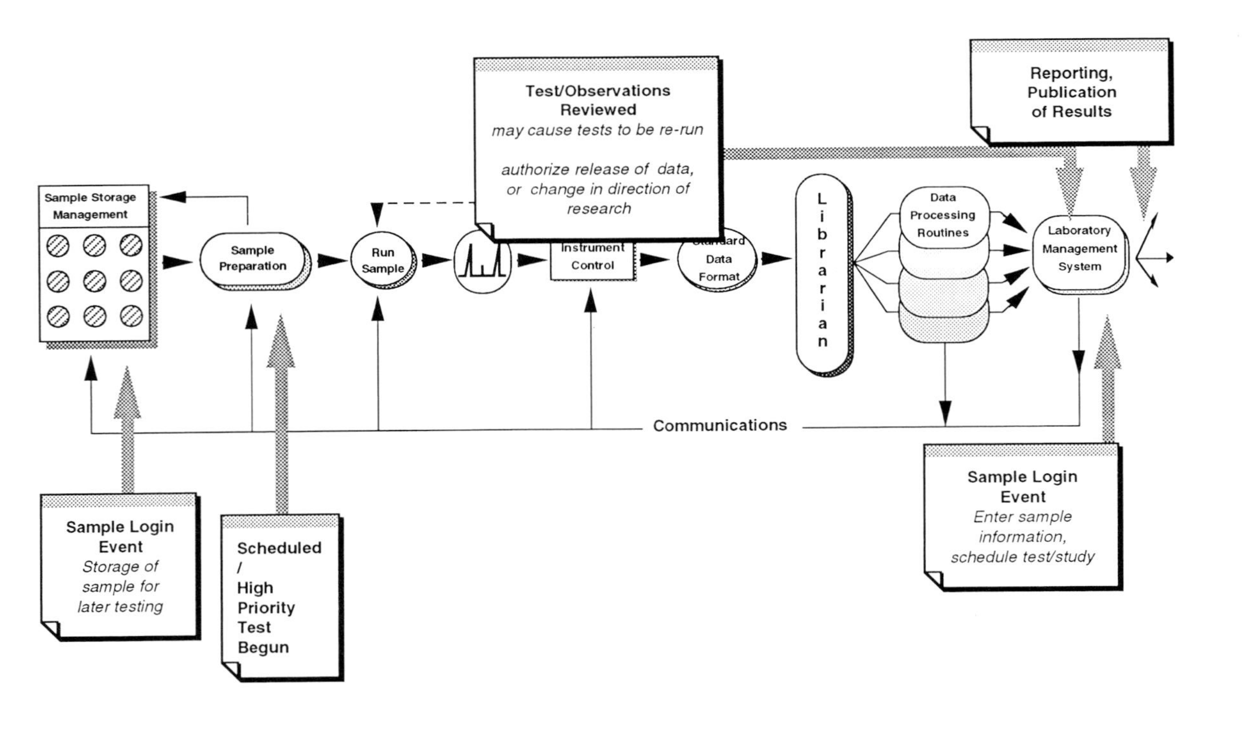


**Figure 12.3**
This figure shows the same process as 12.1, but includes the events that make the process function. This figure is used with permission of the LASF

structure for laboratory operations (showing the relationship between functions). An outline for required standard operating procedures, includes communications between components as part of the system (either electronic or manual), and, helps define the context for a *statement of need.*

The trend in laboratory automation implementation is toward "integrated systems" - following the same direction as commercial applications in office automation. The model shown in Figs. 12.2 and 12.3 can be used as the basis for designing integrated systems. The model is intended to be implemented in a modular fashion. Each module should be validated/qualified as should its communications link both upstream and downstream from it in the process flow. This adds work to the development of a validation plan. Laboratory automation programs in which each element (data acquisition system, robot, LIMS) is treated as an independent entity (no electronic integration), validate/qualify those elements with separate validation programs. The structure (Figs. 12.2 and 12.3) includes that work, plus the need to validate the connections between systems - something that should be required for the communication between any processes whether that communications is electronic or manual.

### 12.4.2 Product Specification and Selection - Determining Product Suitability

The statement of need should include:

- why the need exists
- what is to be accomplished by the system, and how it relates to other systems
- what the characteristics of the system should be, with a statement of which functions are mandatory, and those that are highly desirable (but optional)
- the process being used to determine appropriate vendors and the desirable characteristics of those vendors development and support programs; this should include the design of request-for-quote documents, vendor scoring

procedures[10], etc.

- how the final vendor is going to be determined

- how product short-comings will be addressed (rarely will 100% of all mandatory requirements be met for products as initially offered by a vendor, some customization - particularly on LIMS - will be required)

- training laboratory personnel in the products use

Most of these points are reasonably straightforward, however the fourth consideration does require some comment. Evaluation of the vendors development and support programs is a natural part of evaluating a product, particularly computer systems, since upgrades are frequent, bugs need to be fixed, and products need to be maintained in the light of changes to the underlying operating systems. One way of approaching this issue is through a vendor audit - visiting the vendor and examining his procedures.

The FDA does not require a vendor audit[11]; however it has become an accepted practice and some vendors support visits to their facilities for that purpose. The purpose of the audit to develop confidence that the vendor knows what he is doing in product design and support. This may include discussions with engineering managers to evaluate procedures as well as meetings with the engineers themselves - the details of the meetings should be discussed with the vendor well in advance including expectations for those you want to meet: some vendors may charge for extended meetings particularly if development engineers will be involved. Among the points that needed to be covered is that of problem reporting and resolution - how do you notify the vendor of an operational problem, what response time can you expect, what is the problem escalation process. All of these are measures of the product and a company's commitment to quality.

The document from the UK Pharmaceutical Industry Computer

---

[10]An example of a scoring system for evaluating products can be found in LIMS Workbook, Transition Labs, Golden Colorado, May 1992, section IV page 24

[11]This matter is under FDA review (as of August 1994)

Systems Validation Forum (7] covers this procedure with reference to the European ISO 9000 process: *"For configurable systems, the customer will want to inspect the development process....The audit may find that a supplier has either a well established formal Quality Management System or attained recognized third party Quality Assurance certification such as ISO 9000 (BS5750)"*. The ISO process is not recognized by the FDA, but many vendors use it a symbol of quality in their advertising and presentations. Its application to a validation plan should be considered, however at this point the FDA will not accept it as a substitute for in-house investigations.

## 12.5 Software and Programmable Systems

Fundamentally anything that is programmable has to undergo the same development process. It may occur on the customer's site for custom developed systems, or, within the vendor's engineering department for products. It does not matter if the end results are modules that are compiled at the users site, or firmware in a chip that is part of a hardware package. One part of the ISO 9000 certification for software companies is a matter of showing that a well defined process for product development and support exists, and, that it is followed. A key component of that process is the software development life-cycle [2, 10], widely applied in regulated environments for system design (Fig. 12.4).

The development of software products, for either in-house or commercial production, follows a similar process (in-house developers may short-circuit the process; however that practice is not advisable):

- The development of a **product requirements document**. This is a statement of need. Why is the product being developed, what markets are being served, gross functionality, target hardware platforms, constraints imposed by external factors, and how it relates to other products/system. The support requirements should also be detailed: who is going to support it, plans for upgrades, how will they be phased in, and how many versions will be supported.

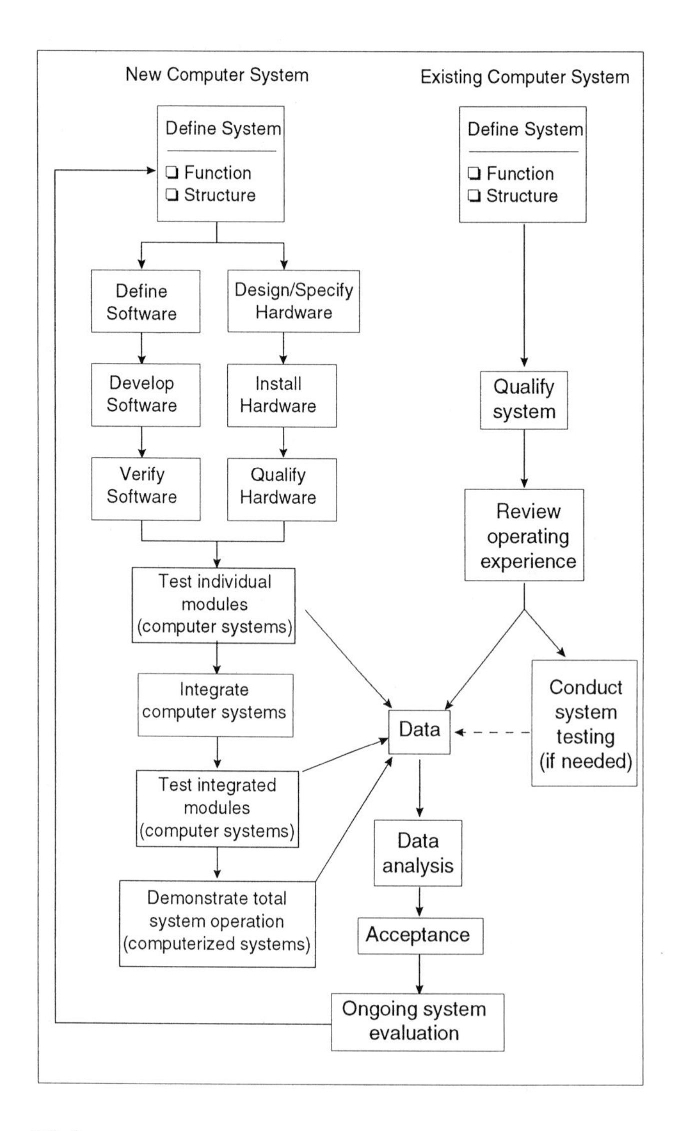


**Figure 12.4**
Computerized systems life-cycle diagram developed by PMA [8], showing the stages in the development and validation of new and existing computerized systems

- Based on the requirements, a **functional specification** is produced. The purpose of this document is to provide in detail, what the product is going to do, how it will work, what capabilities will exist in the initial release, what capabilities will be planned for future versions, and what documentation will be created to support the product (in large systems this should require a separate **documentation plan**).

- In response to the functional specification, two documents are produced. One is the **engineering design specification**. This is how the functions described above will be implemented, specifying tools to be used, operating systems required, hardware required, algorithms used, constraints. The second is the **test plan**. The test plan is an independent document, preferably drafted by someone other than the development engineers, that describes the testing procedures to be used, how each module will be examined, how code reviews will be conducted, and the procedures for documenting bugs, resolving issues and determining the criteria to be met in order for the software to be deemed ready for release into production.

- A **project plan** provides the management details of the project including the provisional schedule (to be refined and updated as the project progress), the people required to do the work, trade-offs between people and schedule, and any contingencies for problems that may be encountered (at as far as they can be foreseen). It should also include the requirements for acceptance testing and completing validation if it is an in-house project or identification of internal testers of the completed product, beta-testers, and the process used to determine when and how each stage of product development is to be reviewed before moving on to the next.

All of these stages are part of the validation plan for in-house projects. When evaluating commercial software, these are the items that should be considered for review during a vendor audit. In the latter case, special arrangements and legal documents may be required; vendors are not likely to expose their inner-most secrets without very good cause and protection. The engineering process may vary from site to site and company to company, but they should have a process and follow it which is part of the ISO 9000 program.

An FDA report [3] notes two types of tests that can be used in evaluating software: functional (black box) and structural (white box). Functional testing treats software as a module and examines the inputs and outputs to determine if it is working properly. Structural testing is a detailed evaluation the source code, examining the logic and flow of control. Every path is examined. Among the points to be evaluated are the quality of the code, its internal documentation, correctness of algorithm implementation, and, looking for dead code (code that is never executed).

This brings us to an interesting point: source code. The FDA requires that users of software systems have the source code available to them for all applications software. Note that at this time, this does not include commercial software in wide-spread use. However, any code that acquires, manipulates, or manages laboratory data should be available for examination. The point of this is to give the users a means of demonstrating that something in the code is not biasing or altering data without it being a conscious decision and part of a Standard Operating Procedure for which they are accountable. Vendors of laboratory software are aware of this requirement, and may either sell a source code license, or, put the source code in an escrow account, giving the right of access under a given set of conditions. The details of these agreements need to be worked out between you and the vendor. Certainly any code that you develop or that is developed for you under contract should have this requirement as part of your engineering practices.

There are tools available to assist developers in maintaining software: code management systems. These are useful since they keep track of current versions of modules as well as the components that went into earlier versions of the software. The use of these systems helps provide a “developer’s audit trail” of programming changes and ensure that the product is built from only the appropriate files.

## 12.6 Vendors and Regulatory Compliance

The marketing potential of the word "validation" has not been lost on laboratory systems vendors. The use of "GLP certified" labels is declining as vendors are realizing that the term is meaningless; they can test components, but since *validation* requires a context (a user-defined need with product requirements) they can not be "pre-validated".

What a vendor can do is significant. They can show that products are well designed, well tested, provide product performance information, and show where they are applicable. They can also train users in product use, provide SOPs for maintenance and maintenance schedules, describe normal operating procedures, provide qualification tests to show that a product is operating as designed (Installation Qualification, Operational Qualification, and Performance Qualification).

Systems that require customization and upgrades, particularly Laboratory Information Management Systems (LIMS) can benefit from the services provided by some vendors. In addition to the points in the previous paragraph, some vendors provide validation test suites. They range from a simple set of instructions of what a person should do in front of a screen to test managers that can do some testing automatically.

One vendor has a program that, during the initial testing phases of a system, "learns" the keystrokes needed to carryout a function and records the results. Repeated testing of that function can be automated by having the computer execute the function and compare the response to what is expected. Deviations are noted and provide a flag to the tester to check that point in more detail. When looking at hundreds of functions and keystrokes, this is a significant benefit. Automated audit trails, software control of revisions to systems, and automated security also provide assistance in meeting regulatory requirements.

Some vendors provide a very useful service to their customers through on-line bulletin boards. These systems allow vendors to provide up-to-date information about systems, software, and an opportunity for users to exchange information about products (including problems). Services like this can be an important part of programs to maintain software.

## 12.7 Training Requirements for Automated Systems

Personnel should be trained properly in the use of automated equipment. That includes an understanding of:

- how to use the system (including applications software, operating systems, how to start the system, proper shut-down procedures, steps to be taken in case of problems),
- system security - how to use, change and protect passwords, and,
- backup procedures.

Participation in training is one obvious mechanism but this alone not sufficient. New people enter a laboratory, existing personnel are exposed to new program features, and the need to refresh people on particular topics are all reasons for the need to provide training outside the traditional classroom. Training people on production systems is dangerous. Mistakes can be serious, and the users will be understandably nervous about the impact of an error.

Having a spare system for training users on new features or refreshing them on existing ones is a good alternative. Some vendors provide for this by allowing you to have additional, non-production copies of a systems available for training purposes.

## 12.8 Special Considerations

The software and computing environment we have today was not envisioned by most laboratory people concerned with systems design and validation. As a result the practices are lagging behind modern technology. There are several areas that warrant additional comment.

### 12.8.1 Emulators

Market pressures to be the "latest-and-greatest" and "all things to all people" have caused computer vendors to find ways of running software designed for one type of computer and operating system to work under their own computing environment; the objective being to provide access to more software. There are two ways of accomplishing this: separate processor boards that contain a second computer or emulators (software/hardware combinations that act like another type of system). The result is that Windows-based products can run on a Macintosh computer, or that Macintosh software can be run on a Sparc-station for example.

The use of these types of applications in laboratories should be done only after very careful consideration. Their primary market is commercial offices and schools, not the real-time world of the scientific laboratory. The original product developers of applications software probably did not intend for their software to be used with emulators and if problems occur they will refer the user to the emulator vendor. The layering of applications upon one O/S that is running under the control of a second O/S, each potentially with its own processor can be cause for two reactions: *admiration for technological juggling,* and, *concern about being able to prove that they system works and is supportable?*

### 12.8.2 Extensions to Operating Systems

Changes to the operating system of computers was once the realm of the vendor or the daring: the user makes changes to the software that controls all the components to a computer system and holds your data. In many cases today it happens without the user being aware of it through the use of INITs (MAC O/S) or TSRs (DOS/Windows). There are a number of sources for these O/S modifiers: anti-viral packages, sound support, type managers, QuickTime, screen savers (some with random animation), adding special features to systems, automatic file compression, and so on. They are usually tested in isolated environments - nothing else running but them and the applications they are designed interact with. That leaves thousands of combinations of these extensions, versions of operating systems, and software that have not been tested together. Many are amusing and keep the working environment interesting, they can also pose a danger to stored data.

A recent article [11] describes how an upgrade to a operating system extension uncovered a flaw that existed in a previous version. That flaw caused software crashes, problems with disk performance and corruption of one users operating system. There were similar problems when Microsoft introduced automated disk compression in version 6.0 of MS DOS.

The decision of what software combinations will reside together on the same laboratory computer should be done with great care. In many cases it is better to purchase several computers and dedicate each to a particular problem, rather than run combinations of packages on the same machine. What a vendor claims can be done in an office environment (*you can run word processing, spreadsheets, data base packages, send and receive FAXes, simultaneously)*, can cripple a data acquisition system. When combinations are run, the validation effort should include the behavior of the system with each combination running to avoid conflicts that may not be apparent when each is run alone.

### 12.8.3 Existing Operating Systems, New Processors

Operating systems are a list of instructions which when executed, carry out a set of functions. That set of instructions assumes that a particular computer is present, since each processor family has its own unique complement of instructions that it can execute. As new processors emerge in a processor family, they add new capabilities in a way that lets the older set still function properly (there are occasional problems: the transition to the 68040 caused some Mac software incompatibilities; in addition some software is optimized for particular versions of machines so the proper software version must be matched to the right machine). The history of use required by retrospective validation for an operating system and commercial software is still largely intact since the new processors should be a proper superset of the old.

When new computer architectures are introduced, that history is no longer applicable. New machines may contain emulator modes that allow older software to run; this is not the same issue noted above, but some caution needs to be exercised. Fitting existing O/Ss on new architectures requires changes to that software. Modules that control device interfaces may have to be rewritten entirely if the O/S has changed. Those changes may have an impact on the behavior of the applications software, so a re-validation phase has to be entered. In general, major releases of operating systems, migration to new hardware platforms, and the introduction of operating system add-ons, are all reasons to consider system re-validation.

### 12.8.4 Heuristic Systems and Programming

These paragraphs are essentially the equivalent of a caution sign, recognizing that this is an area that needs attention from both a theoretical and practical perspective. Almost all the work that has been done in validation has been done on static systems. Their programming and behavior changes in a controlled fashion, under programmer control, with ample documentation and testing.

How does one validate something that may change on a hourly basis or perhaps as a result of the testing procedure itself (neural network system are one example)? Unless the ability to change is turned-off and the static system evaluated, one does not have a stable computing environment with which to work. Unless the system is frozen at some point - and as a result no longer heuristic - conventional validation techniques may not apply. Retrospective validation may apply but that is questionable. The use of simulators is another possibility. We do not have any answers, but the question is an interesting one to consider.

### 12.8.5 Data Exchange/Updating through Application Links

The major operating system vendors have developed data exchange/updating linkages between applications: Publish/Subscribe on the Mac, Dynamic Data Exchange in Windows, and, Hot Links in VMS Workstations, are just three examples. They provide a means of moving and updating information between applications without the need to cut-and-paste or manually enter the information; their automatic modes prevent one from forgetting to do an update should something change. This saves time, effort, and reduces errors in data transfer. It also executes these functions without the benefit of automatic audit trails or any record that changes have occurred, which is the crux of the problems in meeting regulatory requirements.

All regulatory agencies require that changes to recorded information be done in an orderly process, capturing the earlier data, who changed it, why, when, and who authorized it. The mechanism noted does not provide this facility and it may be possible for changes to occur without the user being aware of it, depending on the implementation within a given commercial or user-developed application. The use of these programming facilities needs to be done with care, avoiding automatic updates, and at least maintaining a written record of changes.

### 12.8.6 System Security and Backup

For completeness, laboratory data systems require protection from unauthorized access and data loss. Passwords are the key ingredient in maintaining system security, both on time-sharing systems, lab data acquisition systems, and, desktop machines.

Laboratory personnel need to be trained in the use of passwords to provide controlled access to equipment. Many regard their use as a obstacle to getting work done since they must log-off (or be automatically logged-off) and then log-in to use the system which appears to waste time. Passwords are needed if you are going to ensure that someone does not accidentally or intentionally use a system and lose/alter data through an error or other means. This includes access through the keyboard as well as network access.

SOPs need to be developed for changing passwords on a regular basis, making sure that they are really changed (some count the execution of the password changing program as a change without verifying that the same password is not being used repeatedly), and that the passwords are protected (not written on a blotter or calendar). In addition SOPs are required to gain remote access to systems. Who can dial-in, are call-backs required, what happens if repeated access failures occur?

Backup procedures are another issue. Who is responsible for backup? Are they trained? Is adequate equipment provided for backup? Don't expect people to regularly backup a 200 MB+ disk onto 1 MB floppies - it won't happen. Provide tapes or other backup media that permit a backup procedure to occur quickly and with little intervention. If the process is a problem it either will not be followed or an error will occur that can result in data loss.

The backup process should be tested. Just because a procedure is completed without a reported error, does not mean that everything you need was backed-up; only the things it was told to backup were addressed. There should have enough material on backup media to completely rebuild the system to its current state. That means that the backup software has to be on an easily loadable medium and that diskettes or other bootable media exist to run the system. Having all of the data on tape is fine. Make sure that bootable versions of the operating system exist on bootable media, along with the software necessary to read and operate the tape drive and recovery system.

## 12.9 The Validation of a Simple System

The following is an example, which is included to provide a useful illustration of the validation of a computerized system. A laboratory is doing a gravimetric analysis. The only measuring instrument is a balance. In order to avoid transcription and calculation errors, and to provide a basis for integrating the data into a LIMS, they are looking for an alternative to manual procedures.

### 12.9.1 Statement of Need and Product Requirements

A method is needed to acquire data from a balance during the course of a gravimetric analysis and perform calculations. The results of the calculations and the data used, should be put into a format that can be read by a LIMS without having to manually re-enter any of the data. After careful consideration, it is decided that a computer will be used to capture the data, perform the results, and prepare an ASCII file that can be read by any LIMS system, given a known file format. The product requirements for the software are:

- The program has to be able to read the data from a balance and have the assurance that the readings are correct; provision has to be made for the procedure to work up to 20 samples at a time.

- Calculations described in the SOP have to be performed and reported in printed form and in the specified ASCII file format. The file should contain the sample ID, each weighing (labeled with the step number) and the calculated result.

- The user-interface should be constructed so that the steps that the operator is to take next are clear, errors messages should be clear and in plain text as should recommended corrective action. Errors should be logged to the printer and data files.

- The data acquired at each step should be displayed to the operator. The program should have limits for stage of the measurements to determine if something out of the ordinary is occurring (if a weight loss is expected, then the data should be compared to previous measurements for consistency, if a history file of weighing containers is retained, the tare weights for the containers should be compared to that file to look for anomalies).

- The software should support the communications protocol of the balance - if the balance does not have error checking built in, then each reading should be requested three times and the results compared for consistency.

- Printed documentation should exist to describe the systems operation and use, backup procedures, what to do in case of problems, who to contact for additional support. That documentation should also include the commented/documented source code for the applications software. If the software is developed using a programming language, spreadsheet, or other commercial tools, you should not need the sources for those underlying software packages.

The computer hardware/operating system should be capable of:

- Supporting the software
- Have local storage to load programs and hold data for transmission to the LIMS
- Support communications to the balance and LIMS - over separate communication lines
- Provide passwords to gain access to the system
- Provide backup media sufficient to completely backup the system in one operation

The balance used should:

- be capable of weighing material in ranges specified in the SOP and with the required accuracy and precision
- have a vendor supplied and supported communication port
- have a well documented communications command structure and test procedures for ensuring that the balance's communications is operating properly
- preferably have a built in error correcting communications protocol for transmitting data to the computer; lacking that, have the ability to send the same data repeatedly upon request

These recommednations assume that the work is being done either by an in-house group, or, by an outside.

### 12.9.2 Engineering Response to Requirements

At this point the developers have been told what is required. Now it is time for them to respond with a proposal, including the material noted in section 12.5. Their requirements will either have to be constrained by the existing balances (if they are still qualified for the work, given the addition of a communications requirement) or provide additional technical requirements that will add to the qualification criteria for balances to be considered as appropriate. There may be some iteration between the engineering specifications and the qualification criteria, as the reality of features of available products and the developers technical wish list come into synchronization (developers may wish for a balance supporting Ethernet with a full OSI[12] communications link, they will probably have to settle for RS-232). The changes in specifications should be documented because there is a requirement for an audit trail.

The specifications should contain a list of the qualification criteria for computer hardware and software to be used for the project and the final target production system. In addition any changes to the LIMS system should be determined and a separate engineering schedule should be designed to meet those development requirements, including how the file from the lab data system is going to be transferred to the LIMS and incorporated into its data structure (note: this is the equivalent of *"magic happens here"* in some cartoons, our intent is to outline a project not give the details of its implementation).

Once the specifications and project plan have been agreed to, the equipment needed, including the balance(s) and cables, are selected from the choices available, and qualified for purchase. Once purchased they are tested to ensure that they meet the criteria in the engineering documents. Each step is documented.

The system then goes into the development, testing, and documentation stages until the completed system is ready for evaluation. This time period should also include structural and functional testing.

### 12.9.3 User-Acceptance Evaluation

Evaluation of the delivered system should include the actual system as well as all the supporting documentation (software sources, user manuals, engineering documents, test suites, test reports). All

---

[12] Open Systems Interconnect - refers to ISO 7498 available from American National Standards Association, New York

components should be assembled and put into a simulated production environment for testing. The operators should be trained and then start the evaluation of the complete hardware-software-balance system. The testing includes parallel operation of the same procedures done manually and electronically.

Once the system has been shown to conform to the user requirements and engineering specifications, the system is ready for production use. The current SOPs should be changed to reflect the new procedures because the manual procedure should be included in the revised SOP as a backup to the automated system in case of system failure. Again, all work is documented.

### 12.9.4 Integration with LIMS

Two parallel development paths merge at this point. Their common point of communications is the transmission of a file from the lab data system to the LIMS. Each can be developed and tested independently. The laboratory system is expected to produce a file with a particular format - that can be verified. The LIMS can be tested with files that conform to the expected file format, synthesized based on the specification. Now it must be shown that:

- the output file formats conform to specifications
- the communications path functions properly
- the LIMS can accept and integrate the files from the data system

All of the testing should be done on a LIMS test-bed data base, not the production system. Once the integration testing is completed satisfactorily - the acceptance criteria is defined early in the project - and any additional training is given to the operators, the completed system is put into production (an end-of-project party is recommended, documentation is not required).

### 12.9.5 Project Completion

At this point the user should be able to go back to the original project definition and show a documented path through design, development, testing, integration, acceptance, and entry into production for the entire system. That being the case the user would have a validated system.

## 12.10 Summary

The validation of laboratory computer systems should be considered as a normal part of any project in laboratory automation. It is not just an exercise to meet regulatory requirements, but part of a process of building confidence in the tools and technologies that laboratories depend upon for their existence. It is a matter of adopting an engineering mentality in systems design and implementation. As systems change, the methodologies used to show that they are functioning properly are going to become more sophisticated.

The material covered in this chapter is a good outline for the validation of systems as we understand and apply them today. It will be modified as new hardware and software technologies are introduced into the laboratory, as regulations change, and as we move toward and achieve the levels of integration and functionality that are needed in a competitive and global corporate/economic climate.

## References

1. Guide to Inspection of Computerized Systems in Drug Processing, U.S. Dept. of Health and Human Services, Public Health Service, Food and Drug Administration, Feb. 1983

2. Guideline on General Principles of Process Validation, U.S. Dept. of Health and Human Services, Public Health Service, Food and Drug Administration, May 1987, reprinted May 1990

3. J. Liscouski, *Laboratory and Scientific Computing: A Strategic*

*Approach,* John Wiley & Sons, New York (1994)

4. Technical Report: Software Development Activities, U.S. Dept. of Health and Human Services, Public Health Service, Food and Drug Administration, July 1987

5. 2nd Annual Conference on the Validation of Laboratory Systems, October 6th & 7th, 1994, held in Nashua NH.

6. *InfoWorld*, **17(8)**, 1 (1995)

7. *InfoWorld*, **17(13)**, 1 (1995)

8. R.F. Tetzlaff, *Pharm. Tech.,* **16**, 70-124 (1992)

9. "Pharmaceutical Industry Supplier Guidance: Validation of Automated Systems in Pharmaceutical Manufacture", UK Pharmaceutical Industry Computer Systems Validation Forum, February 1994, Version 1.0 Issue A

10. PMA's Computer Systems Validation Committee, *Pharm.Tech.,* **10**, 24-35 (1986)

11. *InfoWorld*, **16**,, 102 (1994)

## Other References of Interest

J.B. Doherty, *Comp. Intel. Lab. Syst: Lab. Inform. Man.*, **13** , 135 (1991)

S.P Maj, *Comp. Intel. Lab. Syst: Lab. Inform. Man.*, **13**, 157 (1991)

B. Levey and Leonard, *Am. Lab.,* **25**, 54 (1993)

T. Stokes, *Comp. Intel. Lab. Syst: Lab. Inform. Man.* **21**, 131 (1993)

M. Rosser, *Pharm. Tech.* **18**, 74 (1994).

S.H. Segaistad and M.J. Synnevag, *Comp. Intel. Lab. Syst: Lab. Inform. Man.* **26**, 1 (1994)

J.M. Andrade, *Comp. Intel. Lab. Syst: Lab. Inform. Man.* **26**, 13 (1994)

K.G. Chapman, J.R. Harris, A.R. Bluhm, and J.J.Errico, *Pharm. Tech.* **11**,24-35 (1987)

A.S. Clark, *Pharm. Tech,* **12**, 60-66 (1998)

**References that may be of value in software development and testing**

ANSI/IEEE Standard 729-1983. Glossary of Software Engineering Terminology

H. Helms, The McGraw-Hill Computer Handbook. New York: McGraw-Hill Book Company, 1983

G.J. Myers, The Art of Software Testing. New York: Wiley-Interscience, 1979

ANSI/IEEE Standard 730-1984. Software Quality Assurance Plans

ANSI/IEEE Standard 828-1983. Software Configuration Management Plans

ANSI/IEEE Standard 829-1983. Software Test Documentation

ANSI/IEEE Standard 830-1984. Software Requirements Specifications

G.J. Myers, Software Reliability: Principles and Practices, New York, Wiley-Interscience, 1976

R. Dunn and R. Ullman, Quality Assurance for Computer Software, New York, McGraw-Hill, 1982

S. Klim, Systematic Software Testing for Micro Computer Systems, International Planning Information, 1984

# INDEX